Berichte aus dem Institut für Mehrphasenströmung

Band 5

Influence of heterogeneous bubbly flows on mixing and mass transfer performance in stirred tanks for mammalian cell cultivation

- A study in transparent 3 L and 12 000 L reactors -

Vom Promotionsausschuss der Technischen Universität Hamburg
zur Erlangung des akademischen Grades
Doktor-Ingenieur (Dr.-Ing.)

genehmigte Dissertation

von
Annika Rosseburg

aus
Mannheim

2019

Bibliografische Information der Deutschen Nationalbibliothek

Die Deutsche Nationalbibliothek verzeichnet diese Publikation in der
Deutschen Nationalbibliografie; detaillierte bibliographische Daten sind im Internet
über http://dnb.d-nb.de abrufbar.

1. Aufl. - Göttingen: Cuvillier, 2019
    Zugl.: (TU) Hamburg, Univ., Diss., 2019

1. Gutachter: Prof. Dr.-Ing. Michael Schlüter
2. Gutachter: Prof. Dr. rer. nat. Andreas Liese
3. Gutachter: Dr.-Ing. Thomas Wucherpfennig

Vorsitzender des Prüfungsausschuss: Prof. Dr.-Ing. Stefan Heinrich
Tag der mündlichen Prüfung: 23. August 2019

© CUVILLIER VERLAG, Göttingen 2019
    Nonnenstieg 8, 37075 Göttingen
    Telefon: 0551-54724-0
    Telefax: 0551-54724-21
    www.cuvillier.de

1. Auflage, 2019
Gedruckt auf umweltfreundlichem, säurefreiem Papier aus nachhaltiger Forstwirtschaft.

    ISBN 978-3-7369-7108-0
    eISBN 978-3-7369-6108-1

Vorwort

Die vorliegende Arbeit entstand während meiner Tätigkeit als wissenschaftliche Mitarbeiterin am Institut für Mehrphasenströmungen an der Technischen Universität Hamburg unter der Leitung von Herrn Prof. Dr.-Ing. Michael Schlüter. In der Zeit vom Juni 2013 bis Dezember 2018 habe ich das Kooperationsprojekt in Zusammenarbeit mit Boehringer Ingelheim Pharma GmbH & Co.KG bearbeitet, welches die Grundlage dieser Dissertation gebildet hat. Ich bedanke mich bei Boehringer Ingelheim für die finanzielle Förderung.

Besonders möchte ich meinen Doktorvater, Herrn Prof. Dr.-Ing. Michael Schlüter, für die Betreuung dieser Arbeit und für das entgegengebrachte Vertrauen danken. Die vielen fachlichen Diskussionen und die guten Ratschläge haben mich immer wieder aufs Neue motiviert.

Ebenso gilt mein Dank meinen Gutachtern, Herrn Prof. Liese und Herrn Dr.-Ing. Thomas Wucherpfennig sowie Herrn Prof. Dr.-Ing. Stefan Heinrich für die Übernahme des Vorsitz.

Ich danke auch im großen Maße meinen Projektpartnern von Boehringer Ingelheim, für die vielen fachlichen Diskussionen während unserer gemeinsamen Zeit und vor allem für die Möglichkeit ein Teil in diesem einmaligen Projekt gewesen zu sein.

Aber auch meinen Kollegen möchte ich danken, für die gegenseitigen Unterstützungen, den vielen Kaffeerunden und dem einen oder anderen After-Work. Wegen des guten Zusammenhalts und der immer freundlichen Arbeitsweise bin ich jederzeit gerne ins Institut gekommen.

Einen ganz wichtigen Beitrag haben auch die vielen studentischen Arbeiten geleistet, die ich in meiner Zeit als wissenschaftliche Mitarbeiterin betreuen durfte. Ich bedanke mich bei Leonard Parakenings, Christine Klook, Alvaro Miravalles, Khatera Sadat, Birger Niclas, Tim Kuczynski, Daniela Sanders, Vanessa Berg und Sebastian Nichtern. Besonders bedanke ich mich bei Marc Maly für seine Geduld und für die unzähligen Blasen, die er per Hand ausgewertet hat und bei Julián Giraldo-Ospina für seine handwerkliche Unterstützung. Und natürlich bedanke ich mich bei Jürgen Fitschen, der als mein letzter Student mit Angelschnur im Technikum stand und mir später als geschätzter Kollege immer den Rücken frei gehalten hat.

Zuletzt danke ich meinen Freunden und meiner Familie, die mir vor allem in der letzten Zeit immer den nötigen Ausgleich gegeben haben. Danke Svenja, für deine verrückte Art, die mich immer wieder zum Lachen bringt. Danke Simon, dass du immer ein offenes Ohr und eine Schulter zum Anlehnen hast.

Hamburg, im August 2019

Table of content

Nomenclature

Latin

Symbol	Meaning	Unit
A	surface area	m²
A_{tot}	total surface area	m²
a	specific surface area	m² m⁻³
c	concentration	mol m⁻³
c_A	concentration species A	mol m⁻³
$c_{A\infty}$	concentration species A in bulk phase	mol m⁻³
c_{A0}	concentration species A close to wall	mol m⁻³
$\overline{c_A^*}$	mean saturation concentration	mol m⁻³
c_D	drag coefficient	-
D_{AB}	diffusion coefficient	m² s⁻¹
D	tank diameter	m
D	stirrer diameter	m
d	diameter	m
d_b	bubble diameter	m
d_{32}	Sauter mean diameter	m
d_{m3}	volumetric mean diameter	m
do	diameter orifices	m
g	acceleration of gravity	m s⁻²
H	reactor height	m
h	blade width	m
He^{pc}	Henry coefficient	m³ Pa mol⁻¹
k_L	mass transfer coefficient	m s⁻¹
$k_L a$	vol. mass transfer coefficient	s⁻¹
M	torque	N m
N	stirrer frequency	s⁻¹
$\dot{n}_A$	molar flow density	mol s⁻¹ m⁻²
$\dot{N}_A$	molar flow density	mol s⁻¹
n_s	number of impellers	-
P	power input	W
P/V	specific power input	W m⁻³
p	pressure	Pa
q	gas flow rate	m³ s⁻¹

q_P	pumping capacity	$m^3\ s^{-1}$
Q_0	cumulative sum distribution	-
q_0	number distribution	m^{-1}
Q_3	cumulative volume distribution	-
q_3	volume distribution	m^{-1}
s	bottom clearance	m
t	time	s
t_{mix}	mixing time	s
t_c	circulation time	s
t_M	penetration time	s
T	Temperature	K
u_b	rise velocity of bubble	$m\ s^{-1}$
u_K	fluctuation velocity	$m\ s^{-1}$
u_{Tip}	tip speed	$m\ s^{-1}$
u_r	radial liquid velocity	$m\ s^{-1}$
V	(reactor) volume	m^3
V_{fill}	filling volume of reactor	m^3
w_G^0	superficial gas velocity	$m\ s^{-1}$
z	vertical position	m

Greek

Symbol	Meaning	Unit
δ	thickness of boundary layer	m
ε_g	gas hold-up	-
ε_T	specific dissipated energy	$W\ kg^{-1}$
η	dynamic viscosity	Pa s
η_K	eddy length	m
φ	residence time distribution	
ν	kinematic viscosity	$m^2\ s^{-1}$
ρ	density	$kg\ m^{-3}$
σ	surface tension	$N\ m^{-1}$
τ	contact time	s
τ	shear rate	s^{-1}
$\Delta b_{10,90}$	width of bubble size distribution	m

Dimensionless

Symbol	Meaning
$Fl = \dfrac{q}{n \cdot d^3}$	Flow number
$Fr = \dfrac{d \cdot n^2}{g}$	Froude number
$Ne = \dfrac{P}{\rho \cdot n^3 d^5}$	Newton number
$Re_b = \dfrac{u_b \cdot d_b}{\nu_L}$	bubble Reynolds number
$Re = \dfrac{n \cdot d^2}{\nu_L}$	stirrer Reynolds number
$Sc = \dfrac{\nu_L}{D_{AB}}$	Schmidt number
$Sh = \dfrac{k_L \cdot d_b}{D_{AB}}$	Sherwood number
$We = \dfrac{u_o^2 \rho_L}{\sigma}$	Weber number

Indices

Symbol	Meaning
A	species A
b	bubble
c	critical
cal	calculated
CD	complete dispersion
exp	experimental
f	flooding
g	gas/ gaseous phase/ gassed condition
L	liquid phase
mean	mean value
o	orifices
R	recirculation
r	radial
reac	reactor
tot	total
0	unaerated condition
*	saturation condition

Acronyms for stirrers

Symbol	Meaning
BMRF	Bottom mounted agitator
Pb	Pitched blade
Rt	Rushton turbine
2Rt	Two Rushton turbines
Rt-Pb	Rushton pitched blade combination

Abstract

Aerated stirred tank reactors are widely used in chemical industry and bioprocess engineering in which mass transport and mixing time are among the most important parameters for the characterisation and scale-up. However, the prediction of theses parameters is still challenging due to large differences of the hydrodynamic inhomogeneities between lab scale and industrial scale. For a better understanding, a precise measurement of the two phase flow is crucial but requires an optical insight and is thus difficult to realise especially on large scales. To overcome this problem and to be able to perform both global and local measurements, a transparent stirred tank reactor on industrial scale has been erected at the Hamburg University of Technology in cooperation with Boehringer Ingelheim Pharma GmbH & Co.KG.

To identify the differences between lab scale and industrial scale, the different dispersion mechanisms and the resulting bubble size distributions within a laboratory scale reactor (3 L) and a industrial scale reactor (12 000 L) were investigated in this work. Furthermore, the influences of the bubbly flow on the mixing time and the mass transfer performance were identified. The results were compared to literature correlations. It is shown that the dispersion mechanisms are significantly different on both scales, as a result of the small power input in context of mammalian cell cultivation. In the laboratory reactor, the dispersion mainly takes place directly at the sparger, leading to a small bubble size distribution at various power inputs. On industrial scale, the gas phase is dispersed by the vortices behind the stirrer blades and thus strongly depends on the power input. Because of the limitation of the overall power input for mammalian cell cultivation, the minimum stirrer frequency at which sufficient dispersion occurs may be exceeded. This in turn can lead to a wide bubble size distribution with an inhomogeneous gaseous phase and strong buoyancy driven flow. A correlation to estimate the transition between homogeneous and heterogeneous flow regime is presented.

These flow regimes have a particularly strong influence on the mixing time. The mixing times for the homogeneous flow regime are in the same range as the values for the unaerated case. However, the mixing time at the same power input, but for the heterogeneous flow regime, can be reduced up to 80 % due to the strong buoyancy driven flow and the enhanced axial mixing. As a result, the correlations that exist for the aerated mixing time in the literature could not predict the trend sufficiently. A new model that includes heterogeneity is developed. It satisfactorily predicts aerated mixing time, especially in the transition area.

An influence of the heterogeneity on mass transport was also identified. A transition to a heterogeneous flow leads to a sudden decrease of the mass transport due to the relation between the bubble size distribution and the specific surface area. Existing correlations do not take this transition into account and therefore overestimate the mass transport in the heterogeneous flow regime. For an improved correlation, the two flow regimes have to be considered and modelled separately.

Zusammenfassung

Begaste Rührreaktoren finden in der chemischen und der pharmazeutischen Industrie immer noch große Anwendung, wobei der Stofftransport und die Mischzeit zu den wichtigsten Einflussparametern für die Charakterisierung und das Scale-Up gehören. Die Vorhersage dieser Parameter ist jedoch aufgrund der großen Unterschiede der hydrodynamischen Inhomogenitäten zwischen Labor- und Industriemaßstab nach wie vor schwierig. Für ein besseres Verständnis ist die Messung der zweiphasigen Strömung entscheidend, erfordert aber einen optischen Zugang und ist daher insbesondere im großen Maßstab schwer zu realisieren. Um dieses Problem zu lösen und sowohl globale als auch lokale Messungen durchführen zu können, wurde ein transparenter Rührkesselreaktor im industriellen Maßstab an der Technischen Universität Hamburg in Kooperation mit Boehringer Ingelheim Pharma GmbH & Co.KG errichtet.

Um die Unterschiede zwischen Labormaßstab und industriellem Maßstab zu identifizieren, wurden in dieser Arbeit die verschiedenen Dispersionsmechanismen und die daraus resultierende Blasengrößenverteilung in einem 3-L-Reaktor im Labormaßstab und einem 12-m³-Reaktor im industriellen Maßstab untersucht. Darüber hinaus wurden die Einflüsse der Blasenströmung auf die Mischzeit und die Stofftransferleistung identifiziert. Die Ergebnisse wurden mit Literaturkorrelationen verglichen. Es wird gezeigt, dass die Mechanismen der Dispergierung auf beiden Skalen signifikant unterschiedlich sind, was auf den geringen Energieeintrag bei tierischer Zellkultivierung zurückzuführen ist. Im Laborreaktor findet die Dispersion hauptsächlich am Begaser statt, was zu einer engen Blasengrößenverteilung bei verschiedenen Leistungseinträgen führt. Im industriellen Maßstab wird die Gasphase durch die Wirbel hinter den Rührerblättern dispergiert und hängt somit stark vom Leistungseintrag ab. Aufgrund der Limitierung im Leistungseintrag kann es zu einer Unterschreitung der minimalen Drehzahl kommen, bei der die Gasphase komplett dispergiert wird. Dies wiederum kann zu einer breiten Blasengrößenverteilung mit einer inhomogenen Gasphase und einer starken auftriebsgetriebenen Strömung führen. Eine Korrelation zur Abschätzung des Übergangs zwischen homogenem und heterogenem Strömungsverhalten wird in dieser Arbeit präsentiert.

Diese Strömungsregime haben einen besonders starken Einfluss auf die Mischzeit. Die Mischzeiten für das homogene Strömungsregime liegen im gleichen Bereich wie die Werte für den unbegasten Fall. Die Mischzeiten bei gleichem Leistungseintrag, aber im heterogenen Strömungsregime, können aufgrund der starken auftriebsgetriebenen Strömung und der verbesserten axialen Vermischung um bis zu 80 % reduziert werden. Infolgedessen konnten die vorhandenen Korrelationen die Mischzeiten im begasten Zustand nicht korrekt vorhersagen. Unter Berücksichtigung der Heterogenität konnte im Rahmen dieser Arbeit eine neue Korrelation entwickelt werden, die die Mischzeiten auch im Übergangsbereich zuverlässig wiedergibt.

Ein Einfluss der Heterogenität auf den Massentransport wurde ebenfalls identifiziert. Ein Übergang zu einer heterogenen Strömung führt zu einer plötzlichen Abnahme des Stofftransports aufgrund des Zusammenhanges zwischen der Blasengrößenverteilung und der spezifischen Oberfläche. Bestehende Korrelationen berücksichtigen diesen Übergang nicht und überschätzen daher den Stofftransport im heterogenen Strömungsbereich. Für eine verbesserte Korrelation müssen die beiden Strömungsregime separat betrachtet und modelliert werden.

1. Introduction

The pharmaceutical industry is a continuously growing market in which new products are developed and launched every year. Due to the easy handling, the production of these pharmaceuticals is usually carried out in aerated stirred tank reactors with a volume of several cubic meters. However, before the products are launched on the market, the manufacturing processes must undergo strict quality assurance, which often takes place in small reactors with a size of two to five litres. The qualification includes proving that the fermentations on the laboratory scale behave in exactly the same way as on industrial scale. This is often done by adapting the laboratory scale to the large scale, which is a time-consuming and costly undertaking.

The lack of reliable models is primarily based on the fact that investigations on important parameters such as mass transport and mixing time are mainly carried out in small laboratory scales. Only at these scales is it possible to obtain in addition to integral measurements also local information such as bubble size distributions or local flow structures. It has been shown that for the reliable prediction of mass transport as well as for the mixing time, these local data are of great importance. In industrial reactors, these measurements are often not feasible, because of a lack of time and optical access. However, some mechanisms, such as the bubble size distribution or the formation of local flow structures are fundamentally different on various scales. Additionally, many existing correlations have been established for microbial cultivation, working at significantly higher energy inputs and aeration rates compared to mammalian cell cultivation. Both parameters have a significant influence on the process and therefore a transfer of the existing correlations is only possible to a limited extent.

In order to design industrial reactors as well as to adapt laboratory scales for qualification more effectively, it is important to obtain a better understanding on both global and local interrelations for the small and the large scale reactors. To overcome this problem of optical access, a transparent pilot plant on an industrial scale with a volume of $V_{Reac} = 12\,000$ L was set up together with Boehringer Ingelheim Pharma GmbH & Co.KG. The aim of this work is to determine the influence of the local flow structure on mass transfer and mixing time and to identify main differences between laboratory scale and industrial scale. With the help of the transparent reactor, it is now possible to determine a reliable mixing time and to compare the results with existing correlations. The influence of the aeration can also be included in the modelling. In addition, it is possible to determine the bubble size distribution and to identify the influence on the mass transport in order to understand basic mechanisms and to develop more reliable models for mass transport.

2. State of the art

Aerated stirred tank reactors are among the earliest types of reactors, but because of their simple handling and flexibility they are still frequently used. These reactors are particularly common in the pharmaceutical industry, where reliable and safe processes are essential. For a high performance of the cell culture, the reactors need to fulfil different tasks. A good mixing is necessary to provide a homogeneous environment for the cells and to prevent local differences in concentrations of nutrients, pH values or temperatures. Additionally, the mass transfer is of great importance to maintain the oxygen supply in aerobic processes. In this chapter the general background of aerated stirred tank reactors will be given.

For a better understanding of the complex two phase flow, first the characteristics of the unaerated agitation will be presented in chapter 2.1. After that the state of the art of hydrodynamics of two phase flows will be provided.

2.1   Mechanically stirred vessels

In the following a general background on the flow characteristics in agitated tanks for the unaerated case is given. It has been shown that a varity of basic concepts are valid for both the unaerated and the aerated cases.

2.1.1   Equipment for stirred tank reactors

The task of mixing is a wide field ranging from laminar mixing of high viscous liquids to turbulent flows in low viscous media and cannot be achieved with a single stirrer type. Thus, the number of different impeller types is high. Typically they are classified according to the viscosity of the used medium and the main induced flow direction [Zlo03], [Dor13]. In Figure 2-1 the main used stirrer types are listed. On the left side the impellers for media with very low viscosity are presented. Due to the low viscosity of the medium, these agitators are mainly used for turbulent mixing with high stirrer frequencies. The diameters of those impellers are around 1/3 to 1/2 of the reactor diameter. The anchor and the helical ribbon impeller on the right side are usually used in highly viscous media and thus are used for laminar mixing. To provide a good mixing despite the laminar flow, those impellers are often larger than the agitators used for low viscosity liquids. [Zlo03], [Tat91]

In most fermentation processes, for instance with algae, the fermentation broth is a critical factor in the design, as the viscosity increases by several orders of magnitude during the process and often develops a non- Newtonian behaviour. The fermentation broth in mammalian cell cultivation on the other hand, often changes mainly in the surface tension. The density and the viscosity remain in the same range as water. Thus, in those processes often the radial pumping Rushton stirrer and the axial pumping pitched blade or segment stirrer are used [Nie98], [Dor13].

Figure 2-1: Classification of stirrers according to the flow pattern and range of viscosity [Jud76]

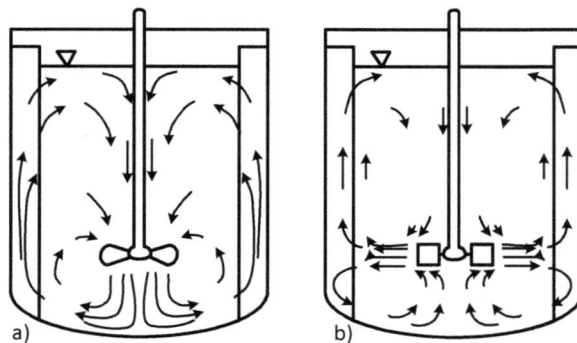

Figure 2-2: Flow pattern on a baffled tank – a) axial flow impeller b) radial flow turbine [Zlo03]

In Figure 2-2 the main induced flow structure of the axial a) and the radial b) pumping impellers can be seen [Zlo03], [Tat91], [Dor13]. Radial pumping impellers such as the Rushton turbine are inducing a high-speed radial flow at the impeller region. At the wall the flow is redirected which leads to an axial flow. The high-speed flow at the discharge of the impeller leads to good dispersion characteristics and makes the Rushton impeller the preferred one when a gaseous phase needs to be dispersed. A disadvantage of this flow behaviour is the tendency to form two compartments, one above and one below the impeller. Within one compartment good mixing can be assumed, but the exchange flow between these compartments is limited. This can lead to a highly increased overall mixing time and is undesirable for industrial processes. [McF96], [Nie98], [Tat91], [Zlo03], [Dor13]

The axial pumping impeller is discharging the material mainly axially but can also lead to a radial flow for a low bottom clearance and large H/D ratio [KW93]. The main tasks of the axial impeller are blending and the dispersion of solids. In fermentation processes they can be found as the second stirrer, mounted above a Rushton turbine, and are used to increase the axial mixing. Furthermore, a downward pumping impeller can also be used to extend holding times of gas bubbles in the system to increase the mass transfer rate [Tat91], [Zlo03], [Dor13].

Additionally to the impeller, further installations such as baffles can manipulate the flow structure in the vessel. In a vessel without baffles and with a central stirrer, the liquid starts to rotate as a solid body as illustrated in Figure 2-3. In this flow regime, the liquid moves like a solid and only poor mixing takes place. For higher stirrer frequencies, the absence of baffles can lead to the formation of a surface vortex that can even reach the impeller with the result of air entrainment. This can lead to high shear forces and is thus undesirable for mammalian cell cultivation. [Tat91], [Zlo03], [Nie98]

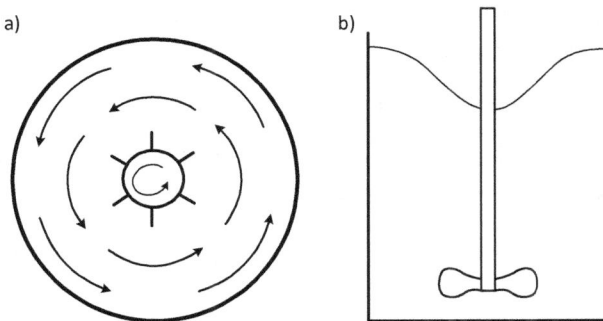

Figure 2-3: **Flow field in an unbaffled agitated tank - a) solid body rotation - b) central surface vortex [Tat91]**

To prevent a solid body rotation and air entrainment, typically wall baffles are installed that disturb the liquid flow and lead to a redirection of the flow. To reach a "complete baffling" of a cylindrical reactor, four baffles are needed that are installed vertically at the reactor wall. [Tat91], [Dor13]

2.1.2 Basic fluid dynamic parameters

The turbulent flow in stirred tank reactors has a high complexity including many different time dependent flow structures that have to be described. These flow structures include the flows at the agitator with the high-speed discharge flow, at the blade wake regions and at the vortex. Additionally, there are the flow at the wall with the corner flows and the flows around the baffles as well as the bulk flows that includes large circulation zones and toroidal vortices. Due to this complexity, the equations of motion cannot be solved. To be able to describe the liquid flow and mixing in a stirred tank reactor the important, mechanisms need to be found that are often described with the help of dimensionless numbers. [Tat91]

The studies within those reactors occur on different scales- (1) the large - or gross scale studies and (2) the local or detailed studies [Tat91]. The gross scale studies include the overall flow pattern and the gross integral flow measurement (e.g. power input [Zlo67a], [Zlo73], [Arm99] or pumping capacity [Cha16]). It is a rather crude technique that assumes that the flow is reproducible in an average way and that temporal and spatial sequences are repeating. The results of these studies are very simple but also very useful for the basic understanding. The more detailed studies often rely on time depending measurements. It is assumed that the flow is very complex and random so that statistical descriptions are necessary. These measurements include the average velocity profiles and turbulent intensities and are often recorded in lab scale reactors (e.g. [Cut66], [CC88], [KW91], [SC95], [OOK08]).

Due to the size of the reactor this study mainly focuses on the gross scale studies, but for a better understanding of the overall flow, structure a short review about the local phenomena will also be given in this chapter. At first, the global flow structures for different impeller types are presented with the help of important dimensionless numbers for the description of the process. After that, the local flow structures close to the impeller blade and within the bulk of the reactor are described.

Power consumption and global flow pattern

To reduce the complexity, dimensionless numbers are used to describe the flow pattern within the agitated tank. The Reynolds number in stirred tank reactors is defined as

$$Re = \frac{n \cdot d^2}{\nu_L} \qquad (2.1)$$

with the stirrer frequency n, the impeller diameter d and the kinematic viscosity of the liquid ν_L. In stirred tanks, the flow is considered to be laminar for Reynolds numbers $Re < 10$ and turbulent for $Re > 2 \cdot 10^4$ [Nie98]. Animal cell cultivations never operate in the laminar region because the viscosity of the liquid is, during the whole process time, in the same range as water, which leads to Reynolds numbers larger than $Re = 10^4$, even for small stirrer frequencies. [Tat91], [Zlo03], [Dor13]

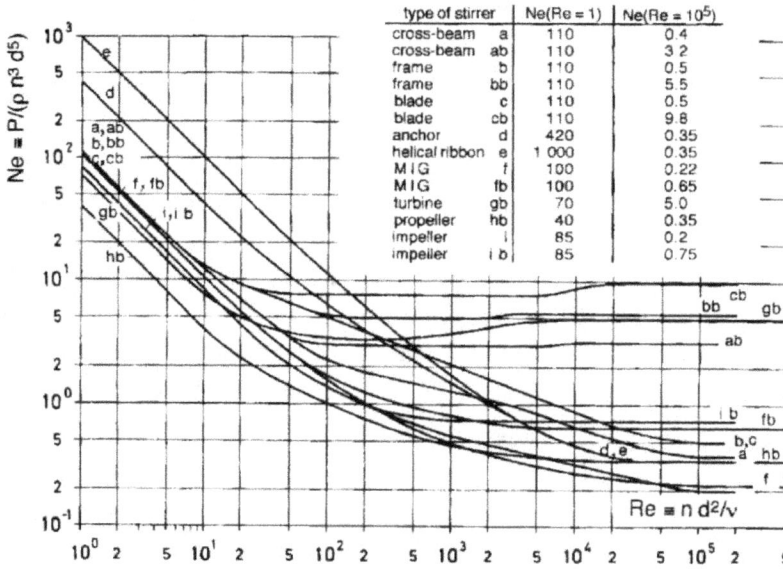

type of stirrer		Ne(Re = 1)	Ne(Re = 10⁵)
cross-beam	a	110	0.4
cross-beam	ab	110	3.2
frame	b	110	0.5
frame	bb	110	5.5
blade	c	110	0.5
blade	cb	110	9.8
anchor	d	420	0.35
helical ribbon	e	1 000	0.35
MIG	f	100	0.22
MIG	fb	100	0.65
turbine	gb	70	5.0
propeller	hb	40	0.35
impeller	i	85	0.2
impeller	i b	85	0.75

Figure 2-4: Newton number for various single impeller types as a function of the Reynolds number (b = baffled) [Zlo67a]

The power input is one of the main characteristics of an impeller. Similar to the drag coefficient for particles or the friction factor for pipes, a dimensionless power input, the Newton number Ne

$$Ne = \frac{P}{\rho \cdot n^3 d^5} \qquad (2.2)$$

can be defined, where P is the power induced to the liquid. For the calculation of the needed motor power, the energy loss in seals and gears also needs to be taken into account. It is thus very difficult to measure the induced power by only measuring the power of the motor. A better way is the determination of the moment M at the stirrer shaft directly below the impeller. With the equation

$$P = 2 \cdot \pi \cdot M \cdot n \qquad (2.3)$$

the induced power can be determined and the Newton number can be calculated. [Tat91], [Zlo03], [Nie98], [Dor13]

The Newton number has already been determined and documented for a list of different stirrers. In Figure 2-4 the Newton numbers for the most widely used types of stirrers are plotted against the Reynolds number. It can be seen that in the laminar region the Newton number is decreasing with increasing Reynolds number. In the turbulent region for $Re > 2 \cdot 10^4$ it can be assumed to be constant. Radial pumping impellers such as the blade impeller ($Ne = 9.8$) or the Rushton turbine ($Ne = 5.0$) have the highest Newton number and are so called high torque impellers. Generally, these impellers need lower stirrer frequencies. Axially pumping impellers generally induce significantly less power at the same stirrer frequency. With a Newton number of $Ne = 0.2$ for the impeller (Figure 2-4 – ib), the induced power is 25 times smaller compared to the Rushton stirrer. Furthermore, the diagram shows that the Newton numbers in systems without baffles decrease considerably more than in baffled systems. [Zlo03], [Dor13], [Bat63], [BNC87]

Besides the power P, the density ρ, the diameter of the stirrer d and the stirrer frequency n, the Newton number is not only dependent on the type of stirrer used in the system but also on combinations of the stirrer and the geometric dimensions of the reactor. The documented values of the Newton number in Figure 2-4 are made for single impeller systems with a ratio $(H/D) = 1$. Due to a better handling, the upscaling to large scale reactors often occurs by increasing the reactor height H instead of increasing the diameter of the reactor [Hir01]. Since the energy dissipation mainly takes place at the impeller region, a second impeller needs to be installed at a higher position to prevent a stagnant zone and obtain a good mixing throughout the reactor [Nie98].

However, this can lead to a dramatic change in the flow pattern and the total power input [Nie98], [Arm99]. For higher clearance of the impeller it can be assumed that the total Newton number is equal to the sum of each single impeller. For other combinations though, the Newton number needs to be estimated separately [Nie98], [Arm99]. There are many more publication that deal with the performance of various stirrer types with various installation conditions. Among others these are the works of Nienow and Miles [Nie71], Gray [Gra82], Rewatkar [Rew10] and also current works from Furukawa [Fur12] and Major-Godlewska and Karcz [Maj18].

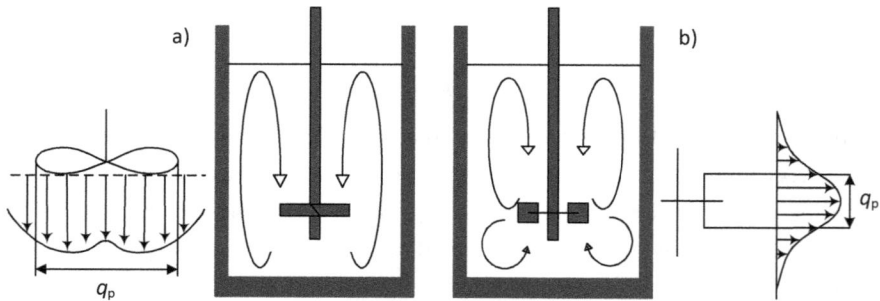

Figure 2-5: Flow pattern of a) a vertical down-pumping pitched blade turbine with a dimensionless pumping capacity of $Fl_p = 0.23$ and b) a radial pumping Rushton turbine with a dimensionless pumping capacity of $Fl_p = 0.24$ [Nie98], [Nie97]

A further characterisation of the hydrodynamics of stirred tank reactors is done with the dimensionless pumping capacity Fl_p

$$Fl_p = \frac{q_p}{n \cdot d^3} \qquad (2.4)$$

with the pumping capacity q_P, which refers to the liquid flow that is pumped by the impeller (see Figure 2-5). The pumping capacity for different stirrer types has been investigated and presented by several authors ([Nie98], [Wu89], [Nie97], [Jaw96], [KW93]). It has been shown that the pumping capacity is almost constant for high Reynolds numbers ([Nie97], [Nie97], [Coo68]) and can be assumed to be $Fl_P = 0.23$ for a down-pumping pitched blade and $Fl_P = 0.24$ for a Rushton turbine.

Local flow fields in stirred tanks

Rushton turbine and blade stirrer

Generally, two different flow characteristics can be found close to the blade; firstly, a very high velocity flow over the blade that can be two times higher than the tip speed of the blades leading to vortices with very high circulation velocities. For a blade or a Rushton turbine two vortices, one at the top and one at the bottom of the blade, are formed. Secondly, a high-speed exit flow occurs that can speed up to 1.3 times of the tip speed of the impeller. This is shown schematically in Figure 2-6. When the jet has left the impeller it is heading towards the wall where it is then redirected and split up into an upward and a downward flow. This redirection leads to the formation of the already mentioned compartments. The vortices on the other hand are turning tangentially after they left the impeller due to the strong Coriolis forces. This turning is schematically shown in Figure 2-7. These two effects of the vortices and the high-speed discharge flow lead to the very high energy dissipation rate in the impeller zone. [Nie98], [CHA81], [VS75]

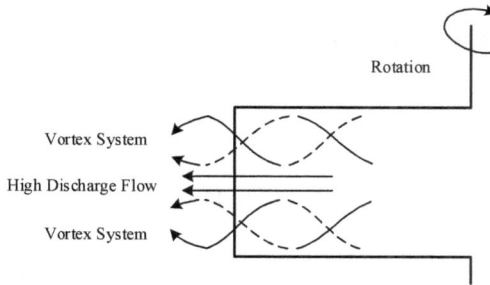

Figure 2-6: **Flow behind a blade with two vortices and a high-speed discharge flow at the side of the stirrer [CHA81]**

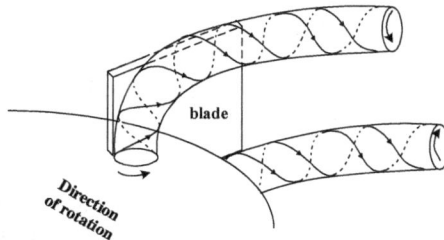

Figure 2-7: **Schematic image of the two vortices behind a stirrer blade [VS75]**

9

Pitched blade turbine

The flow at the impeller of a pitched blade turbine is strongly dependent on the size of the impeller. In small scale systems, the flow at the top and backside of the blades is much higher compared to the front. This leads to vortices that are very chaotic, showing no coherent pattern [Tat80]. In large scale systems for impeller diameters over 0.3 m, the differences in the velocity of the front and the back flow is decreasing and very coherent and strong vortices are formed at the blade tip. For a down-pumping pitched blade in all scales, the main flow intake comes from above and moves slowly towards the blades. Below the blades, a high-speed jet is formed that leads to the typical axial flow behaviour of the pitched blade stirrer. This can be schematically seen in Figure 2-8. [ALI81], [Tat80], [Zlo03], [Dor13]

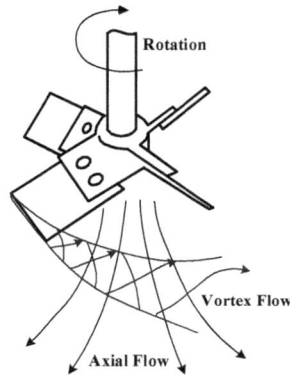

Figure 2-8:   Vortex and axial flow of the pitched blade turbine [Tat91]

General flow

The high-speed discharge flow at the side of the blade leads to a formation of a velocity profile which has been studied intensively. Often the profiles are separated into the tangential, the radial and the vertical velocities. Costes and Couderc [CC88] investigated the radial and the vertical mean flow at different impeller frequencies by using Laser Doppler Anemometry. Rousar [Rou94] and Kresta and Wood [KW91] measured the flow profile which is formed at the side of the blade. They modelled the mean velocity profile and found a similarity between different stirrer sizes. Rousar [Rou94] described the radial mean velocity for the Rushton turbine with

$$\frac{u_r}{u_{tip}} = 0.78 * \exp\left(-\frac{\left(2 \cdot \frac{z}{h}\right)^2}{0.309}\right) \tag{2.5}$$

with the radial velocity u_r, the tip speed of the blade u_{tip}, the vertical position z and the blade height h. The dimensionless velocity profile of equation (2.5) is plotted in Figure 2-9 with experimental data.

Nowadays, the aim is to determine these profiles using computational fluid dynamics. Among these are Ochieng et al. [OOK08], Alcamo et al. and Javed et al. [Jav06]. However, the results are not yet robust enough, so that experimental results are still of great importance

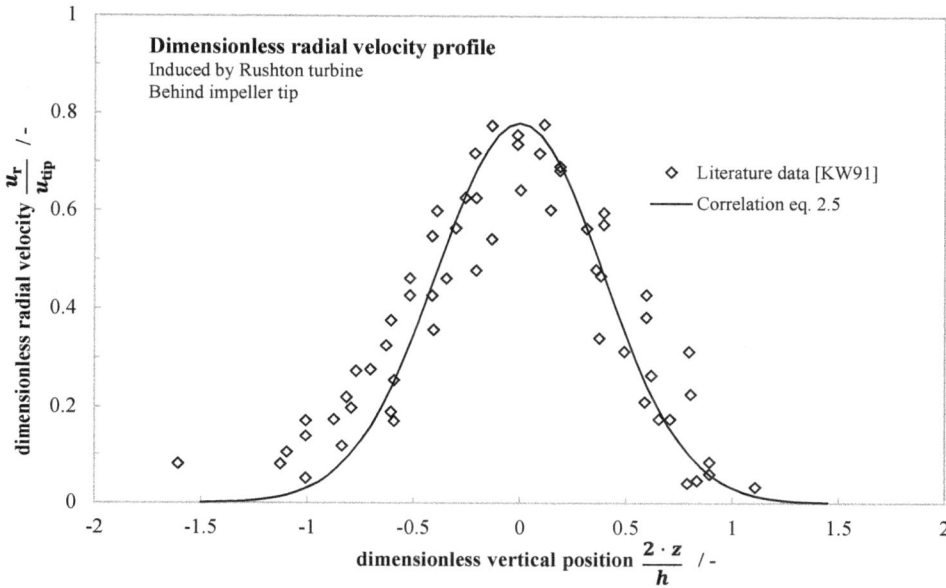

Figure 2-9:   Radial velocity profile [KW91] with experimental data from literature and the proposed correlation eq. (2.5)

2.2   Aerated systems

For the cultivation of mammalian cells, a sufficient supply of oxygen is required. Different techniques are available for the aeration of the fermentation broth. The most common one is the simple aeration with air where a gaseous phase of bubbles with diameters in the range of 100 μm up to several millimeters is present. The dispersion of the gas and the bubbly flow in the bulk phase is a complex system that will be explained briefly in the following section.

2.2.1   Fundamentals of single bubbles

The bubble rise velocity is one of the most critical parameters in aerated systems for the description of the hydrodynamics and the transfer processes [Fan90]. It determines both the gas hold-up and the residence time of the gaseous phase and thus the contact time of the two phases. Furthermore, the boundary layer of the bubble is dependent on the rise velocity and can influence the overall mass transfer.

In industrial applications; a large amount of gas is present; leading to a large amount of gas bubbles in the system which are named a bubble swarm. In a bubble swarm, the shapes and the rise velocities are influenced by the presence of other bubbles. For instance, a higher gas flow rate can lead to a decreased rise velocity, which can be explained by the decreased available flow cross section for the liquid phase [Bra71], [Kra12], [Cli78]. On the other hand, an increase of gas hold-up can lead to more coalescence and thus larger bubbles that rise with a higher rise velocity. Nevertheless, single bubble behaviour is still used to describe the bubble swarms and helps to understand the swarm behaviour.

When a bubble has left the sparger, it starts to rise and is accelerated until the equilibrium between the drag force and the buoyancy force is reached. This leads to the terminal velocity u_b

$$u_b = \sqrt{\frac{4}{3} \frac{d_b \cdot g \cdot \Delta\rho}{c_D \cdot \rho_L}} \tag{2.6}$$

for a single bubble with the gravity g, the density difference $\Delta\rho$, the drag coefficient c_D and the liquid density ρ_L [Bra71].

Besides physical properties and the size of the bubbles, the rise velocity is also dependent on the drag coefficient, which represents the shape and the mobility of the interface. The drag coefficient often is given as a function of the bubble Reynolds number Re_b

$$Re_b = \frac{u_b \cdot d_b}{\nu_L} \tag{2.7}$$

which is the ratio of momentum and characteristic shear stress. In Figure 2-10, the qualitative interrelation between the drag coefficients and the Reynolds number is shown for solid particles, drops and bubbles. Due to the mobile interface between the gas and the liquid and between the liquid-liquid phases, the shape of the bubbles and the droplets are changing at different Reynolds numbers. Hence, different correlations for the determination of the drag force can be found for different Reynolds numbers [Cli78]. Additionally, the different shapes of the bubbles according to Peebles and Gaber [Pee53] are also shown in Figure 2-10 for the different Reynolds number regimes. According to literature, small bubbles behave like a solid sphere with a non-slip boundary (a). When the diameter of the bubble is increasing, the bubbles accelerate and the Reynolds number is increasing. Due to the mobile phase at a certain Reynolds number, the inner volume starts to circulate but the bubble is still formed as a sphere (b). This inner circulation leads to a further reduction of the drag coefficient compared to the solid particle and is stronger for a gas bubble compared to the liquid droplet. With a further increase of the size, the bubble starts to form an ellipsoid which can also be found for droplets (c). This leads to a larger inflow cross-section and thus to a larger drag coefficient compared to a spherical bubble or a solid particle. This form also has a strong impact on the mass transfer. Ellipsoidal bubbles can have up to four times larger mass transfer coefficients. The last condition is an irregular formed bubble (d). Because of their shape, these bubbles are also called umbrella-shaped bubbles. In this regime, the drag coefficient is constant [Pee53]. Due to the different shapes, the drag coefficients are also different in the four different categories. Räbiger and Schlüter give a detailed analysis of the categories that are summarized in [Räb10].

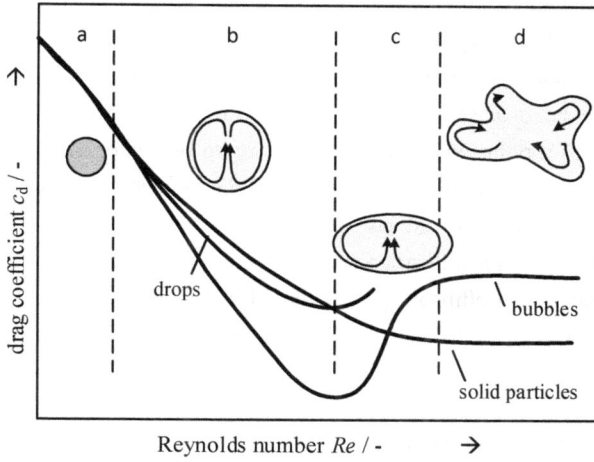

Figure 2-10: Drag coefficient for solid particles, droplets and bubbles as a function of the Reynolds number for the four different bubble shapes [Bot16]

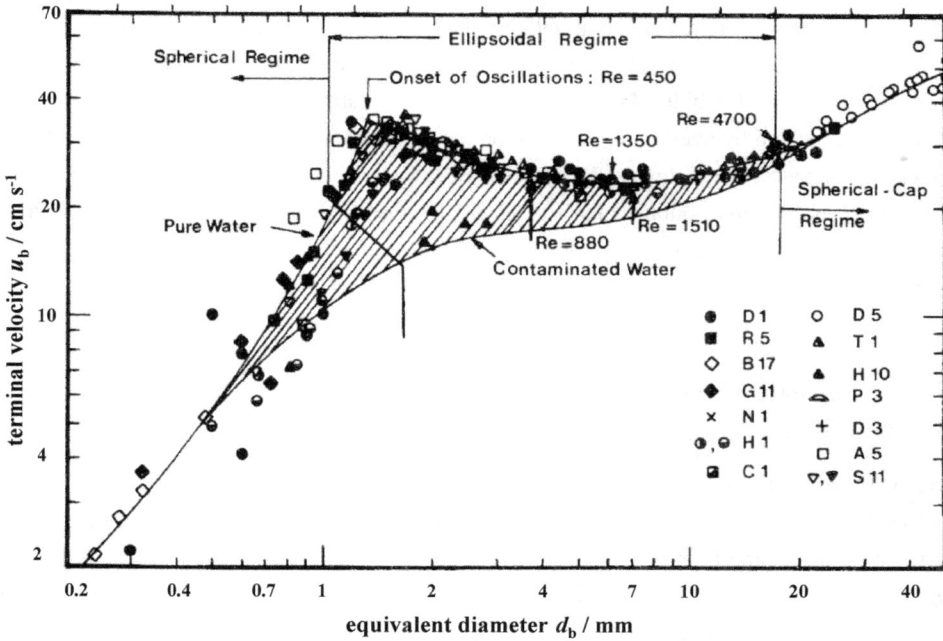

Figure 2-11: Experimental data for the rise velocity of air bubbles in demineralized water, tap water and water with surfactants as a function of the bubble diameter [Cli78]

13

In Figure 2-11 the experimentally determined rise velocity of a single air bubble in water is presented. In this graph, the regimes of the different bubble forms get visible. For small bubbles with a spherical shape, the drag force is low which leads to a low rise velocity. Irregular or umbrella-shaped bubbles have a low drag force which results in a high rise velocity u_b of up to 40 cm/s. For the rise velocity and the shape of a bubble it needs to be distinguished between pure and contaminated systems. In case of a pure system, the rise velocity is increasing with increasing bubble diameter until a point where the bubble starts to oscillate, which leads to a reduction in the rise velocity. In case of a contaminated system this phenomenon cannot be seen. The contamination results in an overall reduction of the rise velocity due to surface active molecules that stabilize the surface of the bubbles. This stabilisation reduces the mobile surface and thus the inner circulation as well as the formation to an ellipsoid gets inhibited. For real systems, this effect is very important, where a lot of surface active substances such as anti-foam or proteins are present.

2.2.2 Fundamentals of bubble swarms

A monodisperse bubble swarm with a uniform distribution over the cross section is only an ideal model system. Due to bubble formation, bubble coalescence and break up, mass transfer and other influences, typically a broad range of bubble sizes is present [Mon08], [Mar08b], [Mar09]. For a better characterisation of the bubble collective, the bubbles are combined in clusters of a certain range of diameter to express the collective with a bubble size distribution. Different mean diameters of a bubble collective can be used for the description. The typical mean diameter d_{mean} is an arithmetic average over all bubbles. For the mass transfer, the surface and the volume of the bubbles are important factors and by averaging, information about one of these gets lost. To prevent this, the Sauter mean diameter has been developed by scientist J. Sauter in 1926 [Wan13]. The Sauter mean diameter is the representative bubble diameter of a collective where the volume and the surface is equal to the original system and can be described with

$$d_{32} = 6 \cdot \frac{V_{g,tot}}{A_{g,tot}} \tag{2.8}$$

where $V_{g,tot}$ is the total volume of the gaseous phase and $A_{g,tot}$ the total surface of the bubble collective [Wan13], [Bou01], [Alv02].

For aerated processes the variety of the bubble size distribution is a criterion of the bubbly flow because the flow regime strongly affects the productivity of a process [Che94], [Bou01]. A bubble column is a typical apparatus of a two phase flow without moving parts and the description of the flow regime can be transferred to stirred tank reactors to some extent. Three different flow regimes can be present in bubble columns: *homogeneous*, *heterogeneous* and *slug flow*. The main fact that influences the transition between the flow regimes is the superficial gas velocity w_g^0

$$w_g^0 = \frac{q}{A_{reac}} \tag{2.9}$$

with the gas volume flow rate q and the cross section of the reactor A_{reac}. The interrelation between the reactor diameter and the superficial gas velocity is illustrated in the flow map in Figure 2-12 [Sha82], [Bou01].

In small diameter columns, often used as laboratory scale reactor, a *slug flow* can occur for very high gas flow rates, when the bubbles start to coalescence. However, this regime is not typical for fermentation processes because an agitator is present and the aeration rate is much lower than in bubble columns [Cam99]. When the bubble sizes vary only in a small range, the bubbly flow can be described as *homogeneous*. In this state the bubbles are mainly rising with the same rise velocity and are homogeneously dispersed over the cross section of the reactor. The horizontal movement of the bubbles and thus coalescence can be neglected. At higher gas flow rates, high shear rates and high pressure forces occur, which affect the bubbles. This leads to coalescence and break-up and thus to a wide bubble size distribution with both small spherical and large irregular-formed bubbles. This flow regime is described as *heterogeneous* flow regime [Sha82] [Che94].

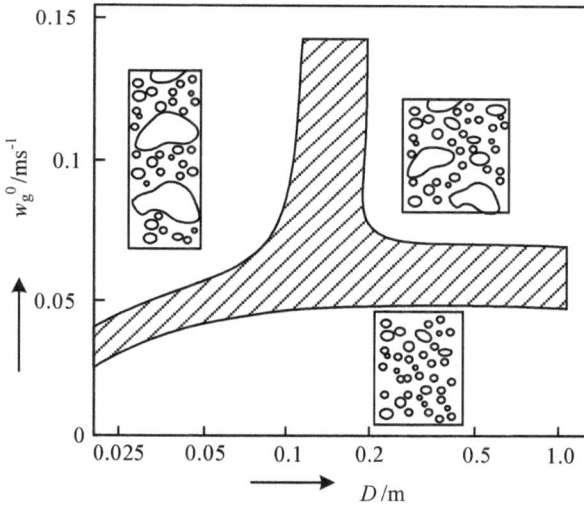

Figure 2-12: Flow regimes in a bubble column as a function of the superficial gas velocity and reactor diameter for a water-air system [Sha82]

Not only the gaseous phase, but also the liquid phase is strongly influenced by the flow regime. This is shown schematically in Figure 2-13. For a homogeneous bubbly flow the gaseous phase is rising homogeneously over the cross section and only small back mixing within a bubble wake occurs. In this flow regime the gaseous phase has only small effects on the global liquid flow. In the heterogeneous flow regime, large umbrella-shaped bubbles are rising with a high velocity and thus lead to a strong movement of the liquid phase. Due to wall effects the large bubbles mainly rise in the centre of the bubble column, which leads to a buoyancy driven flow in the centre and, because of

15

the conservation of mass, to a downward flow at the wall. This downward-directed flow in turn leads to small gas bubbles being trapped within that flow and to a long residence time for these bubbles (see Figure 2-14) [Sha82], [Che94], [Bot16].

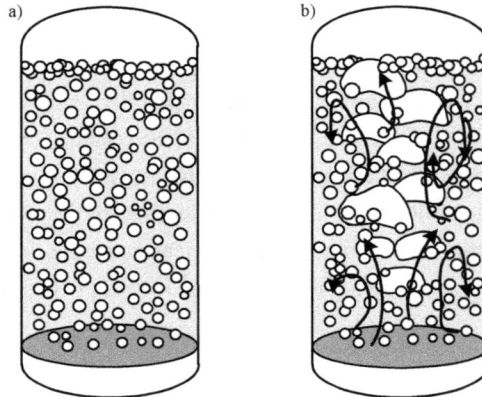

Figure 2-13:   Schematic illustration of a homogeneous (a) and heterogeneous (b) bubbly flow [Che94]

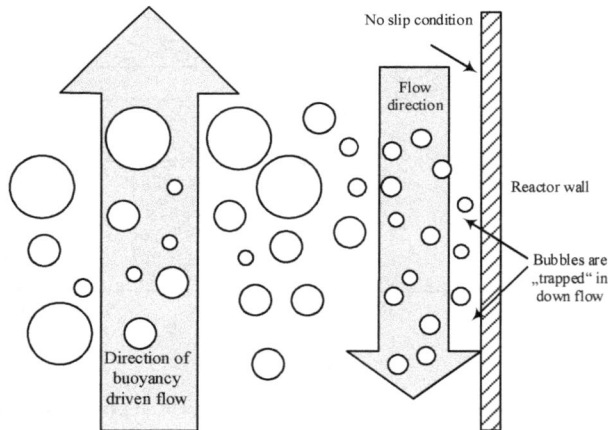

Figure 2-14:   Flow characteristics of a heterogeneous bubbly flow near the wall with a buoyancy driven flow in the middle and the downward flow at the wall of the reactor [Che94]

2.2.3   Gas hold-up and specific surface area

The gas dispersion is often described by the gas hold-up, which is defined as the volume of the gaseous phase in relation to the total volume of the dispersion according to

$$\varepsilon_g = \frac{V_g}{V_{tot}} = \frac{V_g}{V_g + V_L} \approx \frac{V_g}{V_L} \tag{2.10}$$

for an aerated system. A good way to correlate the gas hold-up is to use equations of continuity and motion with population balances. The bubble size distribution, bubble coalescence and break up have to be included, but due to a lack of fundamental knowledge this procedure remains extremely difficult.

The gas hold-up can be estimated by a volume balance over the cross section of the tank. Assuming a constant rise velocity, which is the case for a monodisperse bubble size distribution, the gas hold-up

$$\varepsilon_g = \frac{\sum \pi \cdot d_b^2}{\pi \cdot D^2} = \frac{w_g^0}{u_b} \tag{2.11}$$

can be described as the ratio of the superficial gas velocity to the rise velocity [Tat91].

In the process, the gas hold-up is dominated by the superficial gas velocity but is also dependent on design and operating parameters so it can only be determined by approximation. The relationship between gas hold-up and superficial gas velocity can generally be described as

$$\varepsilon_g \propto \left(w_g^0\right)^n \tag{2.12}.$$

In the homogeneous flow regime, the exponent n is close to one, whereas in the heterogeneous flow regime, when larger bubbles arise, the exponent decreases [Bot16]. Thus, the gas hold-up increases less than proportionally to the gas flow rate [Bot16]. The interrelation of the superficial gas velocity and the gas hold-up can be seen in Figure 2-15.

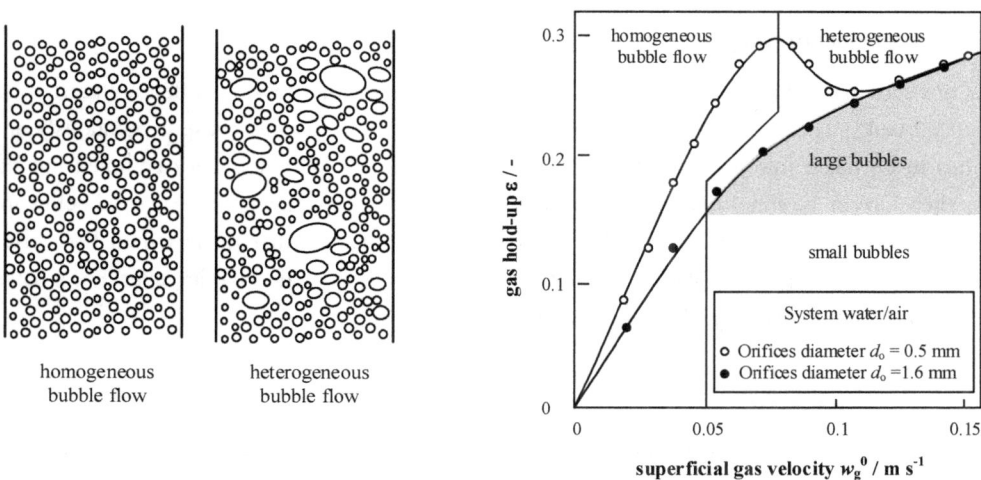

Figure 2-15: Gas hold-up as a function of the superficial gas velocity for two different nozzle diameters [Zah96], [Kri91]

For low superficial gas velocities, the gas hold-up is increasing proportionally. For the case with smaller nozzle diameter, smaller bubbles occur and thus the gas hold-up is higher than with larger nozzle diameter and larger bubbles with a higher rise velocity. For both systems, a homogeneous flow regime can be expected. At the transition to the heterogeneous flow regime the influence of the superficial gas velocity is decreased for both examples. However, for the experiments, with the smaller nozzle diameter, even a sudden drop in the gas hold-up can be seen. This drop of the gas hold-up can be explained by the coalescence of the small bubbles to larger bubbles. The larger bubbles with a much higher rise velocity and shorter residence time result in a smaller gas hold-up. In the heterogeneous flow regime, the exponent of equation (2.12) is decreased to 0.4 to 0.7 depending on the geometrical proportions [Zah96], [Kri91].

Beside of the gas hold-up, the specific surface area a, through which the mass transfer takes place is another important factor. Larger bubbles have a smaller specific surface area in relation to their volumes compared to smaller bubbles. Therefore it is used for the description of the effective contact area between the relevant phases. The specific surface area is defined as

$$a = \frac{A_{tot,b}}{V_{reac}} = \frac{6 \cdot \varepsilon_g}{d_{32}}$$

(2.13)

with the total surface of the gaseous phase divided by the total volume of the liquid phase.

In literature a lot of correlations are available that are describing the gas hold-up as a function of the superficial gas velocity and other important factors. An overview about important correlation for the gas hold-up and the specific surface area in an agitated tank will be given later in chapter 2.2.3.

2.3 Gas dispersion in agitated tanks

So far the behaviour of two phase flows in a reactor w

ithout any moving parts has been presented. However, fermentation processes often are carried out in agitated tanks. The agitation influences both the liquid and the gaseous phase tremendously and also has to be taken into account. The flow behaviour during agitation is very complex but the comprehension can be simplified by the following steps [Tat91]: First, the gas is dispersed below the impeller by the sparger and rises to the impeller where it gets further dispersed in the vortex systems. From there, it starts either to rise to the top or gets recirculated to the impeller. For a better description and comparison, the Flow number Fl

$$Fl = \frac{q}{n \cdot d^3}$$

(2.14)

with the gas flow rate q is used as the dimensionless ratio of the gassing rate to the stirrer frequency. Furthermore, the Froude number Fr

$$Fr = \frac{d \cdot n^2}{g} \tag{2.15},$$

which is the ratio of inertial forces to gravitational forces, is used as the dimensionless stirrer frequency [Tat91], [Zlo03].

For the understanding of the complex phenomena in an aerated stirred tank, it is important to study the flow regime in the bulk phase and the dispersion mechanism in the impeller region. A rough overview of these mechanisms will be described in the following section.

2.3.1 Cavities behind impeller blades

The standard impeller for the dispersion of a gaseous phase is the Rushton turbine, with the sparger below the impeller. Due to the rotation of the stirrer, one or two vortices of high rotating speeds and low pressure occur behind each stirrer blade [Nie98], [Lu89] (see Figure 2-16 and chapter 2.1.2). The gaseous phase gets sucked in by the vortices and leaves them at the end dispersed as small bubbles [Lu89], [Van73]. Their sizes depend on the stirrer frequency and the gas flow rate. At low stirrer frequencies, the buoyancy force of the bubbles dominates over centrifugal forces. In this case, the bubbles do not get trapped by the vortices. At higher stirrer frequencies, the gas is drawn into the vortex system and forms cavities. In this case, the gas dispersion occurs by the vortices and not by eddy dispersion mechanism [Tat91].

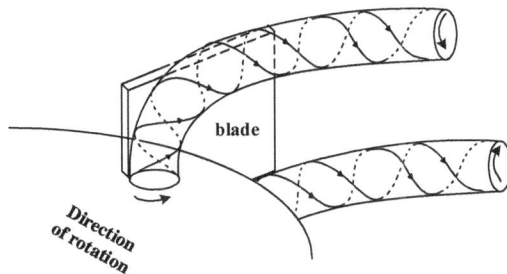

Figure 2-16:   Cavities behind impeller blade [VS75]

Basically three types of cavities can be found (see Figure 2-17), depending on the gas flow rate and on the stirrer frequency [Van73], [Lu89]. For low aeration rates, the so called vortex cavities appear. Here, almost no influence of the gaseous phase on the liquid phase can be found and the overall pattern can be expected as in the single phase. With increasing aeration rate, clinging cavities start to occur. For both the vortex and the clinging cavities, the dispersed gaseous phase is transported radially along the vortices. Large cavities are created at high aeration rates [Lu89], [Lu89]. In these cavities, the centrifugal forces and the vortex are extremely weak and thus the pressure is not low enough to be able to capture the bubbles effectively. For these cavities, the dispersion occurs at the backside of the cavities and due to the weak centrifugal forces, the dispersed gaseous phase is not transported radially as it is the case for the vortex or clinging cavities.

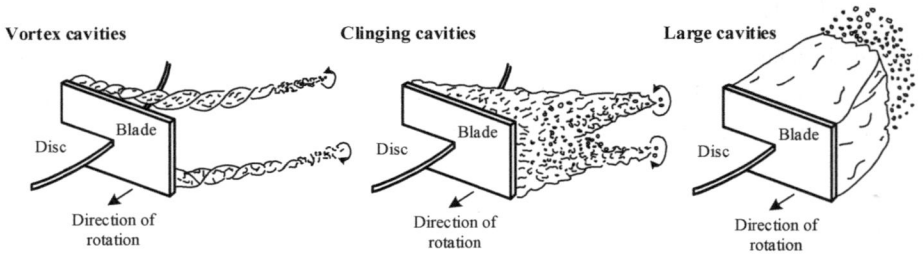

Figure 2-17: Characteristic cavity shapes behind impeller blade at increasing gas flow rate and constant stirrer frequency [Lu89], [War85]

Depending on the stirrer frequency and the gassing rate, different combinations of these cavities have been observed by Nienow et al. in a 6-blade Rushton turbine [Lu89], [War85]. The cavities are schematically illustrated in Figure 2-18. For low gas flow rates, only vortex cavities occur, which start to grow with increasing gas flow rates. At a certain gas flow rate, a combination of three large and three clinging cavities starts to form. This formation is called the "3-3"-structure. It was found that this vortex system is drawing less energy than the other vortex systems. For even higher gas flow rates, ragged cavities form. These cavities are very irregular formed cavities and draw thus much more energy than the 3-3 structure. These cavities form when too much air is present that cannot be dispersed by the impeller. This state is also called "flooding" and will be explained in the following chapter.

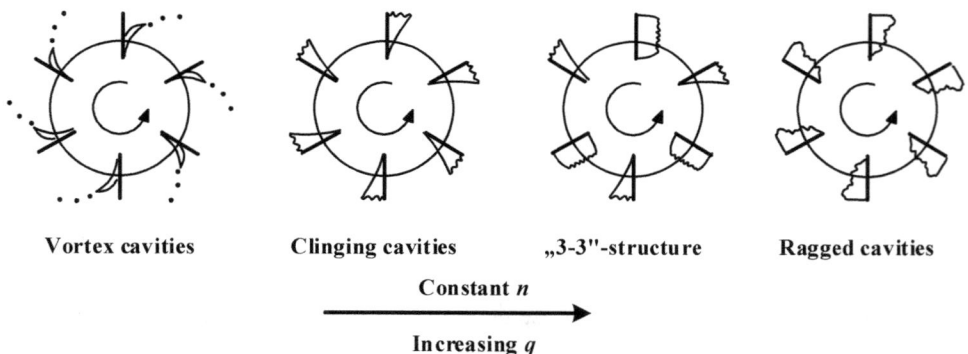

Figure 2-18: Schematic image of the cavity combination with increasing gas flow rates [Nie98], [Lu89]

2.3.2 Specific surface area, gas hold-up and flow regime in aerated stirred tank reactors

Within the bulk phase, typically two regimes can be found: the loading or complete dispersion regime and the flooding regime [Wie83], [Rus68]. Within the loading or complete dispersion regime, the stirrer is able to disperse the gaseous phase, and bubbles are present in the whole reactor. In this regime, the momentum induced by the buoyancy driven flow is lower than the momentum induced by the impeller, so that the bubbles are mainly following the fluid phase (see Figure 2-19) [War85]. For the complete dispersion (Figure 2-19-c), a homogeneous bubbly flow occurs with small bubbles even at the bottom of the reactor below the sparger. In the loading regime (Figure 2-19-b), the gaseous phase still gets dispersed, but the bubbly flow starts to be heterogeneous. Furthermore, fewer bubbles are present in the bottom part of the reactor. When the gassing rate is increased, at some point the impeller is not able to disperse the whole gaseous phase anymore because the cavities behind the impellers are growing until large cavities occur (Figure 2-19-a) ([Nie98], [War85], [Bom06]). At this point, only poor dispersion of the gas is possible as already explained in chapter 2.3.1. As a consequence large bubbles are rising along the stirrer axis, which leads to an undesirable condition. At this point, the momentum induced by the buoyancy driven flow exceeds the momentum induced by the impeller and the main flow pattern is dominated by the upwards rising gaseous phase ([War85], [Rus68]). The same effect can be found when the stirrer frequency is reduced at a constant gassing rate, because the vortices are getting weaker for lower stirrer frequencies.

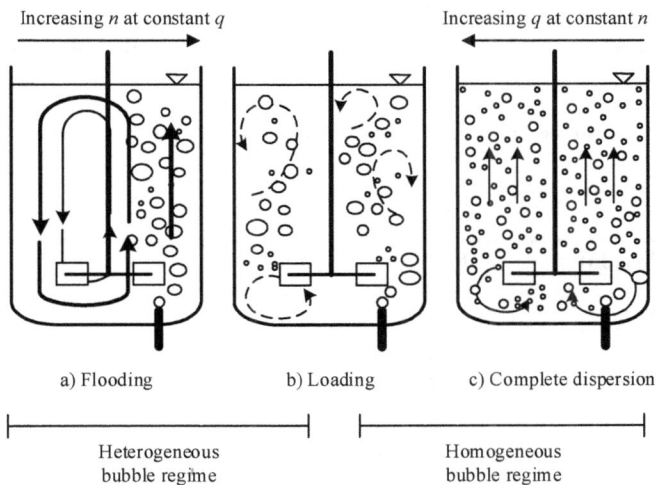

Figure 2-19:   Flow characteristics for different flow regimes

Generally, the loading regime is the preferred operation condition because a large specific surface area is present [Rus68]. Some authors even state that complete dispersion should be aimed for, because gas needs to be present at the bottom of the reactor as well. Otherwise, there is a risk of anaerobic condition in this area. However, these authors ignore the mixing that takes place in the liquid phase [Tat91].

Nevertheless, it is important to know the point of transition where flooding starts to occur. This has been investigated intensively by different authors. Typically, the Flow number is used to describe the aerated condition. Different techniques are presented in literature to determine the transition from loading to flooding. The easiest one is the optical observation but an optical access to the reactor is needed. Additionally, this procedure is very subjective. Van't Riet and Smith [Van73] suggested that the transition can be found by analysing the gas-filled cavities. Another method is the measurement of the power input or the overall gas hold-up ([War85], [Bom06]). Subsequently, for each Froude number Fl, a critical Flow number Fl_c can be defined at which flooding occurs. It has to be mentioned that different techniques can lead to different results for this transition line. Nevertheless, the majority of authors found the critical Flow number to be a function of the Froude number as well as the geometrical factor (D/d) according to

$$Fl_c = a \cdot Fr^b \left(\frac{D}{d}\right)^c \tag{2.16}$$

were a, b and c are fitted constants ([Nie98], [War85], [Bom06], [Pag02]). With Equation (2.16), the maximum gas flow rate at a fixed stirrer frequency can be estimated. Table 2-1 gives a short overview of correlations for a Rushton turbine in water/air. The differences in the correlation become apparent in Figure 2-20, where the results are compared for $D/d = 3$ and Froude numbers of $0.01 < Fr < 10$. It can be seen that the correlation by Zlokarnik [Zlo67b] gives 100 times higher critical Flow numbers than the correlation by Zwieting [Wie83]. While Mikulcova presented in 1967 a range in in which the transition occurs.

Table 2-1:   Correlations for the critical Flow number Fl_c as function of the Froude number Fr in gassed stirred tanks

Author		Formula	Comment	
Zwietering,	[Wie83]	$Fl_c = (D/d)^{-3,3}$		(2.17)
Dickery,	[Wie83]	$Fl_c = 1.88\ Fr^{\,0,5}(D/d)^{-3,3}$	$(D/d) = 3$ $(h/d) = 0.5$	(2.18)
Zlokarnik,	[Zlo67b]	$Fl_c = 1.94\ Fr^{0,75}$	$(D/d) = 3.33$	(2.19)
Mikulcova,	[Wie83]	$Fl_c = 100/(6.11 \pm 1.24)\ Fr\ (D/d)^{-3,43}$		(2.20)

Wiedmann et al. explained these differences in the investigated scales and presented results for three different vessels with the same geometrical ratios $(D/d = 3$ and $h/d = 0.5)$ but different reactor diameters. Their results are presented in Figure 2-21, where the critical Flow number Fl_c is shown for different vessel dimensions in dependency of the Froude number. It can be seen that the larger vessel with a diameter of $D = 1.5$ m reaches higher Flow numbers before flooding occurs than the smaller vessel with a diameter of $D = 0.2$ m. It also becomes clear that the influence of size decreases with increasing vessel diameter.

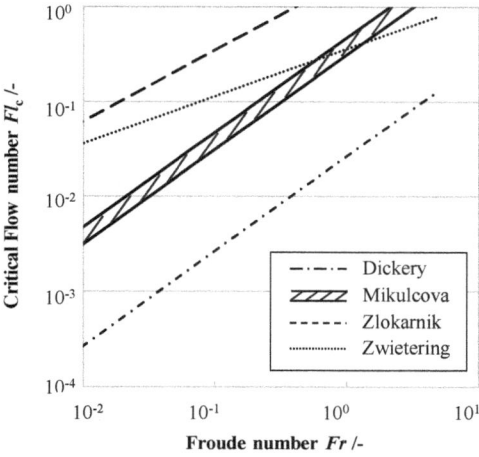

Figure 2-20: Different correlations for the critical Flow number as a function of the Froude number in gassed stirred tanks [Wie83]

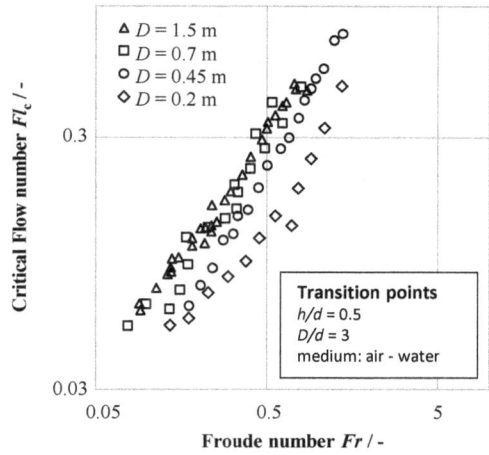

Figure 2-21: Critical Flow number as function of the Froude number for different vessel diameters [Bom97]

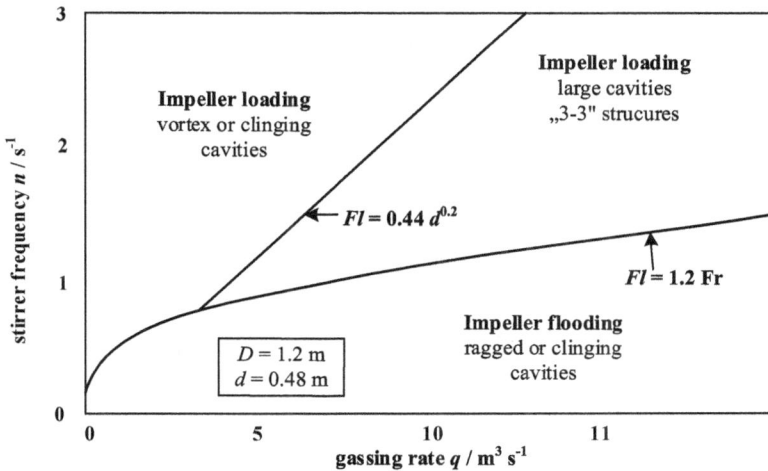

Figure 2-22: Flow map for a vessel with a diameter $D = 1.2$ m and one Rushton turbine [War85]

The transition between loading and complete dispersion regime has also been studied by several authors. Different correlations in the form of equation (2.16) are available to describe the transition between these regimes, for instance by Nienow in [Nie98] or by Lee and Dudukovic [Lee14].

23

Usually, these correlations are combined in a flow map in order to obtain the transition regions as a function of stirrer frequency and gassing rate. Figure 2-22 was presented by Warmokerken and Smith in 1985 for a Rushton turbine with $d = 0.48$ m and a tank diameter $D = 1.2$ m with the included correlation for the flooding/loading and the loading/complete dispersion transition.

Due to the different flow regimes, it is difficult to calculate the gas hold-up and the specific surface area for a wide range of stirrer frequencies and gas flow rates. Since the flooding regime is the unwanted regime, only a few investigations have been undertaken to describe this flow behaviour. For the loading and complete dispersion regime on the other hand, a large number of correlations can be found in literature to describe the gas hold-up ε and the specific surface area a.

Gas hold-up

In chapter 2.2.3, the interrelation between the superficial gas velocity and the gas hold-up has been described for the heterogeneous and the homogeneous flow regime in a non-agitated bubble column. It has been shown that the influence of the superficial gas velocity on the gas hold-up depends on the flow regime. For the homogeneous flow regime, the gas hold-up is strongly dependent on the superficial gas velocity. In the heterogeneous flow regime, the influence is reduced. This different behaviour within the various flow regimes can also be found in the stirred tank reactor, when the gas hold-up is plotted as a function of the Froude number for a constant superficial gas velocity.

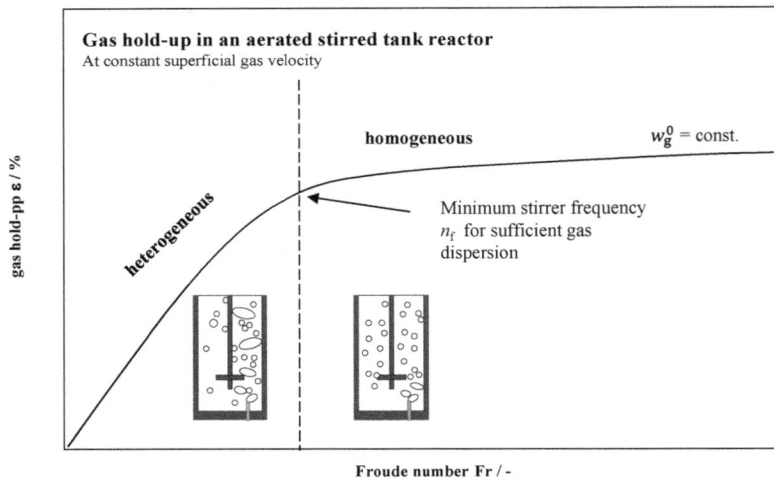

Figure 2-23: Gas hold-up in an agitated stirred tank reactor as a function of the Froude number at constant superficial gas velocity in the heterogeneous and the homogeneous flow regime

This can be schematically seen in Figure 2-23. For low stirrer frequencies, a heterogeneous bubbly flow with a small gas hold-up is present. With increasing stirrer frequency, the gaseous phase gets dispersed and the gas hold-up increases. At a certain point, a further increase of the stirrer frequency does not lead to a further dispersion of the gaseous phase, and consequently no further significant increase of the gas hold-up occurs. At this point the flow regime changes from a heterogeneous to a homogeneous bubbly flow.

A lot of investigations have been conducted to describe the gas hold-up as a function of the process and geometrical parameter. In an early work, Rushton (1944) [Fou44] published a correlation in the general form of

$$\varepsilon_g = c \cdot (n \cdot d)^a \cdot (w_g^0)^b \tag{2.21}$$

with $a = 0.47$ and $b = 0.53$ (eq. (2.22)) [Zlo03]. Further investigations have been performed by Bouaifi and Roustan (1998), Yawalker (2002), Moucha et al. (2003) and more recent by Gao et al. 2001 and can be found in Table 2-2. The values a and b for different correlations vary between 0.2 and 0.7 [Pau04]. Greaves and Bariou (1990) published a correlation in a similar form with the addition of the geometrical ratio (d/D). Smith (1992) and Rewatkar and Joshi (1993) proposed that the gas hold-up can be estimated with dimensionless groups. Calderbank (1958) and Hughmark (1980) included physical properties such as the density and the surface tension, as well as other geometrical properties into the correlation.

Table 2-2:   **List of gas hold-up correlation**

Author	Correlation	System	
Rushton, [Fou44]	$\varepsilon_G \sim \left(\dfrac{P}{V}\right)^{0.4} \cdot (w_g^0)^{0.53}$		(2.22)
Bouaifi and Roustan, [Bou97]	$\varepsilon_G \sim \left(\dfrac{P}{V}\right)^{0.24} \cdot (w_g^0)^{0.65}$		(2.23)
Gao et al., [Smi04]	$\varepsilon_G = 0.9 \left(\dfrac{P}{V \cdot \rho}\right)^{0.2} \cdot (w_g^0)^{0.55}$		(2.24)
Yawalkar, [Yaw02]	$\varepsilon_G = 0.556 \, (\varepsilon_T)^{0.25} \cdot (w_g^0)^{0.041}$	$D = 0.21 - 3.33$ m	(2.25)
Moucha et al., [Mou03]	$\varepsilon_G = 0.017 \left(\dfrac{P}{V}\right)^{0.61} \cdot (w_g^0)^{0.57}$	$D = 0.29$ m Rt	(2.26)
Moucha et al., [Mou03]	$\varepsilon_G = 0.05 \left(\dfrac{P}{V}\right)^{0.49} \cdot (w_g^0)^{0.58}$	$D = 0.29$ m 2RT	(2.27)

Author	Correlation	System	
Greaves and Barigou, [Gre90]	$\varepsilon_G = 4.07 \cdot n^{0.62} \cdot q^{0.64} \cdot \left(\dfrac{d}{D}\right)^{1.39}$	$D = 1$ m Rt	(2.28)
Smith, [Smi92]	$\varepsilon_G = 0.85 \, (Re \cdot Fr \cdot Fl)^{0.35} \cdot \left(\dfrac{d}{D}\right)^{1.25}$	$D = 0.44 - 2.7$ m Rt	(2.29)
Rewatkar and Joshi , [Rew93]	$\varepsilon_G = 3.54 \left(\dfrac{d}{D}\right)^{2.08} Fr^{0.51} \cdot (Fl)^{0.43}$	$D = 0.57$ and 1.5 m Rt	(2.30)
Calderbank (1958), [Bus13]	$\varepsilon_G = \left(w_g^0 \cdot \dfrac{\varepsilon_T}{u_b}\right)^{0.5} + 0.0216 \left[\dfrac{\left(\frac{P}{V}\right)^{0.4} \rho^{0.2}}{\sigma^{0.6}}\right] \left(\dfrac{w_g^0}{u_b}\right)^{0.5}$		(2.31)
Hughmark, [Hug80]	$\varepsilon_G = 0.74 \left(\dfrac{q}{n \cdot V}\right)^{0.5} \left(\dfrac{n^2 d}{g} \dfrac{d^3}{w \cdot V^{\frac{2}{3}}}\right)^{\frac{1}{2}} \left(\dfrac{d_b n^2 d^4}{\sigma \cdot V^{\frac{2}{3}}}\right)^{1/4}$	$D = 0.25 - 1$ m Rt	(2.32)

The overview of the correlations provides an impression of the large number and wide range for an application of the calculation methods. However, no mentioned correlation takes into account the change in the flow structure when flooding occurs. Therefore, it can be assumed that the presented correlations can only be used in the loading and complete dispersion regime. Few authors actually distinguished between different flow regimes of the bulk or the impeller region. Smith and Warmoeskerken [Tat91] distinguished between the impeller flow regimes before the appearance of the 3-3 structure (flooding regime), whre the gas hold-up equals

$$\varepsilon_g = 0.62 \, q^{0.42} n^{1.6} \qquad (2.33)$$

and during the appearance of the 3-3 structure (loading regime), where the gas hold-up equals

$$\varepsilon_g = 0.67 \, q^{0.75} n^{0.7} \qquad (2.34)$$

with the gas flow rate and the stirrer frequency as the only parameter. It can be seen that for higher gas flow rates, just before flooding occurs, the influence of the stirrer frequency is decreased. The influence of the gas flow rate on the other hand is higher for the homogeneous flow regime.

Specific surface area

In contrast to the correlation for the gas hold-up only, few equations that determine the specific surface area are available. This is due to the complex and difficult determination of the surface area.

The determination is often carried out by measuring the gas hold-up and the Sauter mean diameter and subsequently using the formula

$$a = \frac{6 \cdot \varepsilon_g}{d_{32}} \qquad (2.35)$$

for the calculation of the specific surface area [Gar04]. However, some empirical correlations are availabe, such as

$$a \sim (n \cdot d)^{1.1} \cdot (w_g^0)^{0.75} \qquad (2.36)$$

from Yoshida and Miura [Yos63]. This correlation was determined chemically for a 16-blade impeller for different reactor diameters from $D = 0.25$ m to $D = 0.59$ m.

Another, less empirical and also well-known formula is the correlation of Calderbank [Cal58]

$$a = 1.44 \left[\frac{(P/V)^{0.4} \rho_L^{0.2}}{\sigma^{0.6}} \right] \left(\frac{w_g^0}{u_b} \right)^{0.5} \qquad (2.37)$$

that includes the specific power input (P/V), the density of the liquid phase ρ_L, the surface tension σ and the terminal velocity of the bubbles u_b.

2.3.3 Power input in aerated agitated tanks

The power input P_0 in agitated systems has already been described in chapter 2.1.2. However, the gaseous phase can have a strong influence on the overall power input, requiring additional consideration of the gassed power input P_g. Generally, the power input is reduced when a gaseous phase is present [Nie98], [Zlo03], [Tat91]. This can mainly be explained with the pressure drop from the front of the blades to the cavities. This pressure drop becomes smaller when cavities are growing. For very high gas hold-up it is also possible that the decrease of the overall density leads to a decrease in power input, but with small gas hold-ups less than 1 % this only plays a minor role in mammalian cell cultivation.

Often, an additional Newton number Ne_g is used for the aerated processes, and can be described as a function of the Flow number. Figure 2-24 illustrates the Newton number under aerated condition for a Rushton turbine and a pitched blade turbine, determined in a reactor with a diameter of $D = 0.45$ m. Zlokarnik found that the Newton number for a Rushton turbine can fall from $Ne = 5$ for Flow numbers $Fl < 0.001$ down to $Ne_g = 0.5$ at Flow number $Fl > 0.5$ [Zlo67a]. The reduction of the Newton number for a pitched blade turbine has not been studied intensively because they are generally not used for gas dispersion. Furthermore, these impellers do not form cavities like those occurring behind the blades of e.g. Rushton stirrers. Only one vortex per blade is forming at the blade tip [Tat91]. Generally, the reduction of the power input when a gaseous phase is present is much less compared to a blade stirrer. [Jud76].

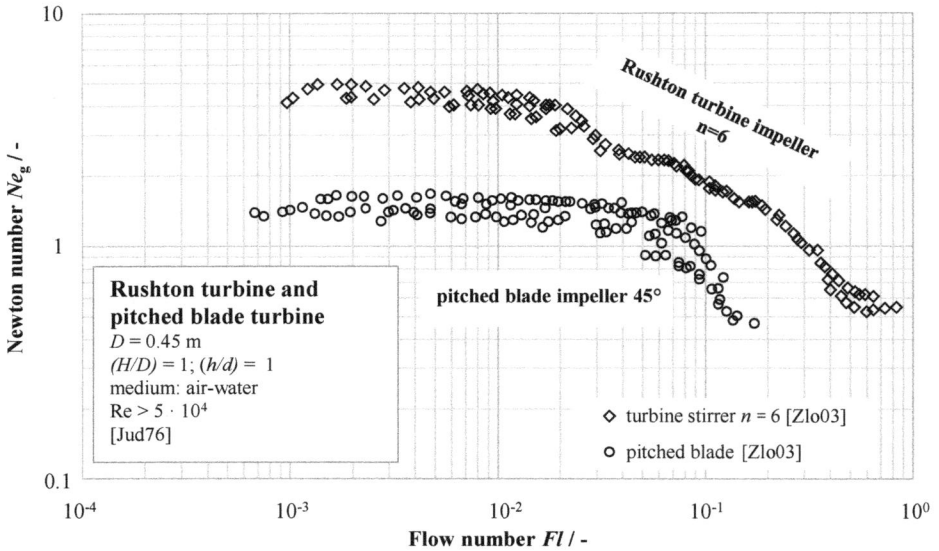

Figure 2-24: Newton number Ne_g under aerated condition for a Rushton stirrer [Jud76]

However, not only an increase in aeration rate can lead to a reduction of the Newton number. With constant gassing, an increase of the stirrer frequency can also result in a decreased power input. This phenomenon can be explained with an increasing gas recirculation with increasing stirrer frequency. Thus, on the basis of the recorded Newton numbers under aerated condition, the different flow regimes can be distinguished from each other. This has been studied for instance by Kapic and Heindel [Kap06]. In Figure 2-25, the Newton number is presented as a function of the Flow number for constant gassing rates. It can be seen that with increasing stirrer frequency the Newton number is slightly increasing until the critical stirrer frequency n_f is reached. Until this point, the impeller is flooded. In the loading regime, an increase of the stirrer frequency leads to a decrease in the Newton number. This decrease can be explained by an increased gas circulation which leads to a higher gas hold-up for a constant gassing rate. This increased gas hold-up in turn leads to growing cavities and to a reduction of the Newton number. When the gaseous phase gets dispersed completely, the cavities behind the blades are shrinking and the Newton number is increasing again. This point ($n = n_{CD}$) is the optimal operating speed for a given gas flow rate because at this point the required energy is the lowest with simultaneous complete dispersion [Kap06].

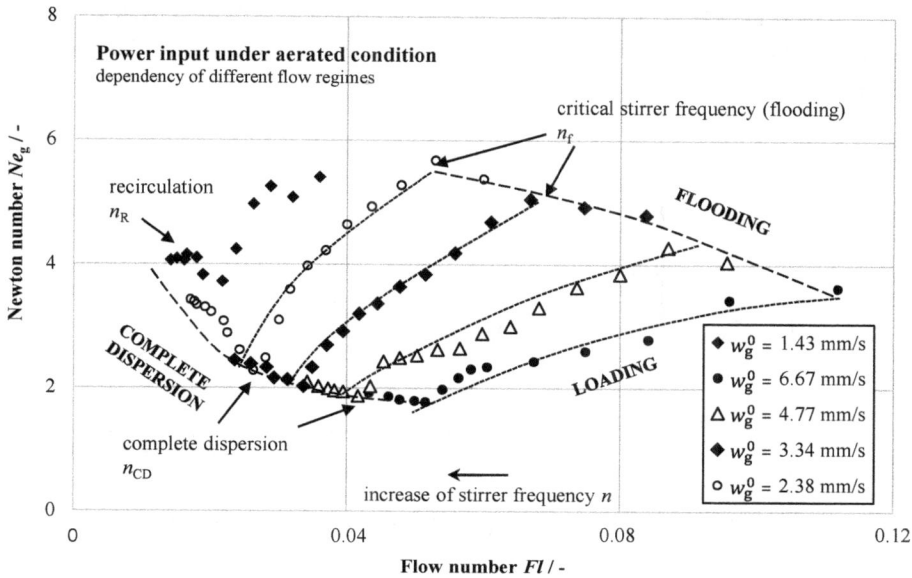

Figure 2-25: Newton number Ne_g under aerated condition as a function of the Flow number at constant superficial gas velocities [Kap06]

2.4 Mixing time in stirred tank reactors

The content of this subchapter correspond in large parts to the published article [Ros18].

The mixing time is a crucial parameter for the characterisation of stirred tank reactors. For instance, during the process of fermentation, fluids are added at regular intervals to the reactor to regulate the pH value or to provide nutrients for the cells. In this case a good mixing behaviour is required to minimize local concentration gradients [War85].

The mixing time is defined as the time that is necessary to achieve a certain degree of homogenisation after the addition of a tracer pulse into the reactor ([Asc15], [Nie97]). Generally, mixing occurs on three different scales. These are the global mixing, the local mixing and the mixing on molecular scale. The global mixing dependents on the global flow pattern of the liquid that can be induced by the agitator or by the buoyancy driven flow induced by a bubble swarm. The local mixing, dependents on the local flow structure such as the small eddies or the back mixing behind a rising bubble. Mixing on molecular scale depends on the molecular diffusion.

Ultimately, mixing is only fully achieved by molecular diffusion. Nevertheless, the time until total mixing occurs can be strongly reduced by the hydrodynamic conditions. In this thesis, only the global and local mixing will be taken into account in this work.

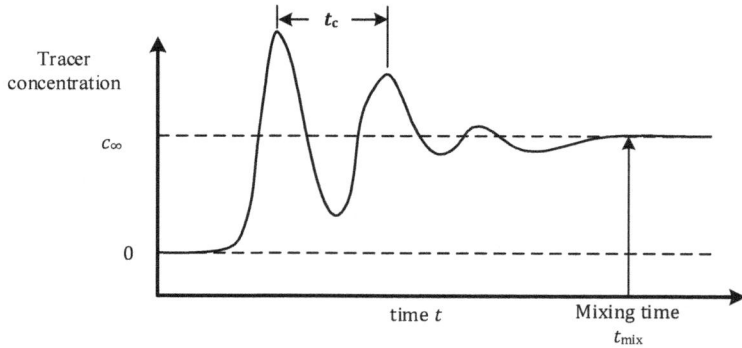

Figure 2-26: **Mixing time and circulation time t_c from a transient response after a pulse of tracer injection [Nie97]**

Two principal measurement techniques are common to determine the mixing time: the first method is a local measurement where a fixed probe is used that measures the concentration of a tracer which is added to the reactor. For instance, conductivity probes are used to measure the ion concentration of a salt that is mixed into an aqueous phase. When the signal reaches a certain threshold of the end value of conductivity, the mixing is defined as completed [Nie97]. Figure 2-26 shows the typical signal for a homogenisation experiment with a conductivity probe. From such a graph, the defined circulation time t_c as well as the mixing time t_{mix} can be determined. The disadvantage of this method is that the mixing time can only be measured locally so no dead zones or flow structures can be detected.

The second method is the decolouration of a dissolved dye, for instance phenolphthalein. When a base is added to the aqueous phenolphthalein solution, the shift in pH value will lead to a change in colour from transparent to pink. By adding an acid afterwards, the pH value is shifted backwards and the medium turns colourless again. The mixing time is defined as the time until the last colour within the reactor disappears. With this method, the global flow structure as well as dead zones can be visualized and measured quantitatively. But a total optical access to the reactor is needed which is often difficult to realise for reactors on industrial scale. In addition to that, concentration gradients and local overshoots of the base and acid concentration cannot be detected with this method [Nie97].

2.4.1 Single phase mixing

To calculate or model the mixing time, a deep understanding of the liquid flow pattern is required. The flow pattern is mainly influenced by the impeller type as well as the gas flow rate. A lot of investigations have been conducted to characterise the flow field of single phase systems for various impeller and reactor types. Zlokarnik investigated the mixing time t_{mix} for single stirrers intensively by using the dimensionless mixing time $t_{mix} \cdot n$ [Zlo67b]. In Figure 2-27 the trends in dependency of the Reynolds number for three single stirrer types are shown. Zlokarnik divided the mixing characteristic into three different areas and showed for the blade stirrer ($d/D = 0.33$; $H/D = 1$) that the dimensionless mixing time is proportional to $Re^{-0.7}$ for $10^1 < Re < 10^3$. In the range of $10^3 < Re < 10^4$, the dimensionless mixing time can be assumed to be almost constant and for the range of $Re > 10^4$ the dimensionless mixing time is proportional to $Re^{0.7}$.

For different stirrer types, the transition between these regimes might change but always three different slopes of the curve can be found. Zlokarnik stated that measurements in large scale reactors are indispensable to detect the third zone correctly because for $Re \gg 10^4$ the mixing time becomes almost zero in lab scale reactors.

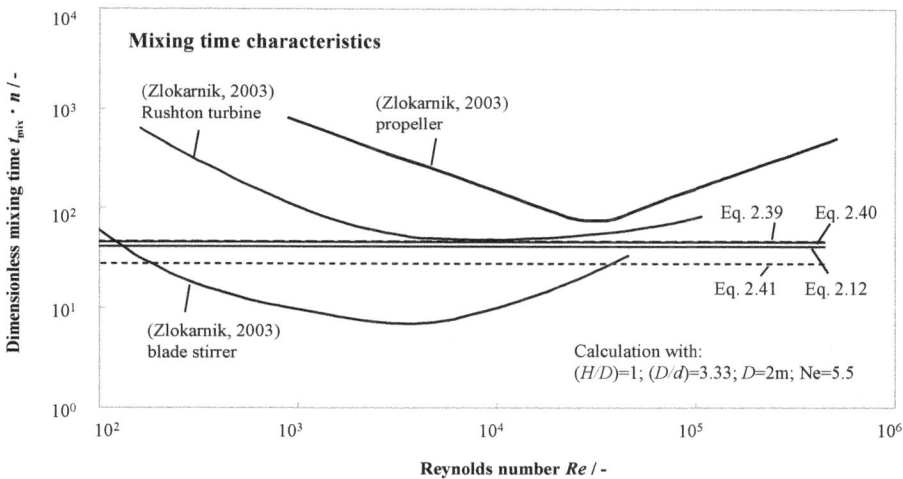

Figure 2-27: Mixing time characteristics for Rushton turbines, propeller stirrers and blade stirrers. All stirrers with the ratios $d/D = 0.33$; $H/D = 1$ [Zlo03]

Further investigations have been conducted for single phase mixing to develop more precise correlations. Different assumptions have been made to derive correlations for the mixing time. Some of them are presented in Table 2-3. For instance, eq. (2.38) is based on the circulation time t_c and assumes that after five circulation loops, the mixing is already completed. The correlation only depends on the ratio of (d/D) and the pumping capacity Fl_p. Other correlations are based on the energy dissipation rate ε_T (eq. (2.40)) or are empirical correlations based on the compartment model such as from Vasconcelos [Vas95] (eq. (2.41)). In contrast to the investigations made by Zlokarnik [Zlo03], all correlations assume that the dimensionless mixing time is independent of the Reynolds number.

31

Most of the investigations on mixing have been performed in a single stirrer system with a ratio of $(H/D) = 1$. Larger reactors, however, are often built much thinner, with larger (H/D) ratios. In order to achieve the desired mixing, additional stirrers are used for larger level ratios, which can lead to completely different flow behaviour. In general, it can be assumed that additional stirrers lead to a shorter mixing time. However, it also needs to be considered that, especially with radially pumping stirrers, compartments may form which in consequence lead to a worse mixing time. Thus, it is important to investigate further into multiple impeller mixing. But only few researches dealt with those systems. Taking a look at correlations (2.41) and (2.42) presented in Table 2-3, with n_s as the numbers of impeller stages, it can be seen that with these correlations the mixing time is reduced with an increase in stages. It is doubtful that they are applicable for any kind of stirrer type.

Table 2-3: Overview of relevant mixing time correlations for stirred tank reactors with the unaerated dimensionless mixing time $t_{mix} \cdot n$

	Author	Formula	Comments	
Single-phase	Nienow (1990), [Nie97]	$t_{mix}^0 \cdot n = 3.9 \left(\frac{d}{D}\right)^{-3} Fl_p^{-1}$		(2.38)
	Van't Riet (1991), [Nie97]	$t_{mix}^0 \cdot n = 3 \left(\frac{d}{D}\right)^{-3} Ne^{-1/3}$		(2.39)
	Grenville (1995), [Nie97]	$t_{mix}^0 \cdot n = 5.9 \, D^{\frac{2}{3}} (\overline{\varepsilon_T})^{-\frac{1}{2}} \left(\frac{d}{D}\right)^{-1/3}$		(2.40)
	Vasconcelos, [Vas95]	$t_{mix}^0 \cdot n = 2.3 \exp(0.68\frac{D}{d} + 0.83 \, n_s)$	With n_s = numbers of impeller	(2.41)
	Sieblist, [Sie16]	$t_{mix}^0 \cdot n = A \left(\frac{D}{d} \cdot \varepsilon_T\right)^{-\frac{1}{3}} \left(\frac{H}{D}\right)^{\frac{2}{3}} (n_s + 1)^{2/3}$	A = fitting constant	(2.42)

In fermentation, additional complexities occur due to the rise of the liquid height. Biological processes are often performed as fed batch processes which results in an increase of the liquid height during the fermentation time. At the beginning, with a low liquid height, the upper impeller is placed close to the surface so that a good mixing can be expected. During the process, the liquid height is rising and the influence of the impeller in the upper region is decreasing. At the end, it can be assumed that two compartments form within the reactor- a well-mixed area in the impeller region and a poor mixed area at the top of the liquid height.

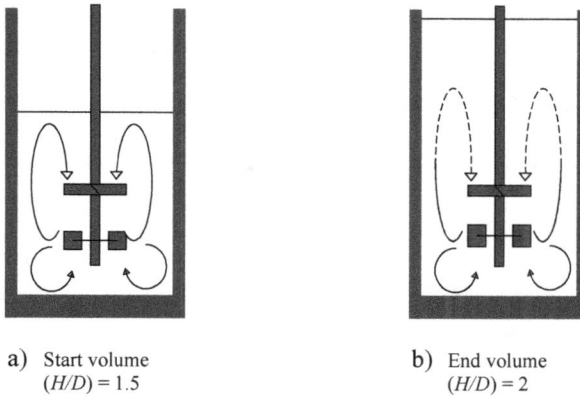

a) Start volume
   $(H/D) = 1.5$

b) End volume
   $(H/D) = 2$

Figure 2-28:   Flow structure for the fermentation at different production times and different filling volumes

2.4.2   Two phase mixing

Fermentations in bioreactors mostly take place under aerated conditions and therefore the buoyancy driven flow induced by the gaseous phase needs to be taken into account for the calculation of mixing time. To describe the influence of the gaseous phase, the Flow number is used. Besides the investigation on the flow regime, a lot of investigations have been undertaken during the last decades on the influence of the gas flow rate on the mixing time. It can be seen in Table 2-4 and Table 2-5 that most of the investigations have been conducted in small scale systems under loading conditions. Nienow [Nie97] and Saito et al. [Sai92] showed that within this regime the aeration has a negative effect on the mixing time. Vasconcelos et al. [Vas95] showed in more detail that this is only the case when the gaseous phase reduces the total energy input of the stirring. Shewale and Pandit [She06] and Alves and Vasconcelos [Alv95] also included the flooding regime into the discussion and showed that the aeration can have both positive and negative effects on the mixing time depending on the flow regime within the reactor. Shewale and Pandit [She06] stated that for a large power input the gaseous phase is distributed well and follows the impeller induced flow. Due to the reduction of the power input in the presence of a gaseous phase, the flow pattern is weakened which results in an increased mixing time compared to the unaerated case.

On the other hand, for a small power input, Shewale and Pandit [She06] stated that the power induced by the gaseous phase is larger than the reduction of the power input of the stirrer. Due to the rise of the bubbles, the mixing between two adjacent compartments is increased which results in a shorter global mixing time compared to the unaerated system.

Table 2-4: Investigation of the influence of the gaseous phase on the mixing time in small scale reactors

Author	Volume / L	Stirrer type	Reynolds number	(P/V) / Wm^{-3}	Gassing / vvm	Method	Comment Infl. of aeration
Haß and Nienow, [Haß89]	150	Rt	$1.3 \cdot 10^5$ - $6.5 \cdot 10^5$	20-500	0-1.5	De-colouration	Loading Increased mixing
Saito, [Sai92]	180	Rt	$6 \cdot 10^4$ - $3 \cdot 10^5$	50-200	0-1	De-colouration	Loading Increased mixing
Vasconcelos, [Nie13] [Vas95]	60-260	2Rt	$3 \cdot 10^4$ - $1 \cdot 10^5$	10-2000	0-0.6	Local measure-ment	Loading Decreased mixing
Alves, [Alv95]	60	3Rt	$3 \cdot 10^4$ - $1 \cdot 10^5$	50-2000		Local measure-ment	Flooding Decreased mixing
Nienow, [Nie96]	70	Rt	$5.4 \cdot 10^4$ - $8 \cdot 10^4$	2-50	0-0.01	De-colouration	Loading Increased mixing
Machon, [Mac00]	75	4Rt	$2 \cdot 10^4$ - $6 \cdot 10^4$	20-500	0.2-0.5	Local measure-ment	Loading/ Flooding Both effects
Shewale, [She06]	70	3Rt	$8 \cdot 10^4$	10-1000	0.3-1.5		Loading/ Flooding
Gabelle, [Gab11]	40-350	2Rt	$4 \cdot 10^4$ - $2 \cdot 10^5$	100-1000	0.4		Loading Decreased mixing
Montante, [Mon15]	12	Rt-Pb	$1 \cdot 10^4$ - $5 \cdot 10^4$	2-250	0.5-2	Local measure-ment	Loading/ Flooding Decreased mixing

Investigations in large scale systems with a filling volume over $V_{fill} = 5$ m³ (Table 2-5) are also available but are mostly conducted under conditions for microbial fermentation with much higher energy input over $(P/V) = 100$ W/m³ and gas flow rates over 1 vvm ([Vrá00], [Vrá99]). Vrábel et al. have validated a compartment model with reactors of $V_{fill} = 8$ -22 m³ for the loading regime. For gas flow rates in the same range as used for mammalian cell cultivation but much higher energy input, they did not find a strong influence of the gaseous phase on the mixing time. Nienow et al. [Nie96] also investigated the mixing time in large scale systems ($V_{fill} = 2$-8m³) with an energy input of $(P/V) = 5$-50 W/m³ and a gas flow rate up to $q = 0.01$ vvm. They showed that the gaseous phase leads to an improvement of the mixing time of up to 50 % and explained it with the buoyancy driven flow of the gaseous phase.

Table 2-5:   Investigation of the influence of the gaseous phase on the mixing time in large scale reactors

Author	Vol. /L	Stirrer type	Reynolds number	(P/V) / W m⁻³	Gassing rate / vvm	Methode	Comment Inf. of aeration
Nienow, [Nie96]	2000-8000	Rt	$1.1 \cdot 10^5$ - $3 \cdot 10^5$	2.0 - 50	0 - 0.01	Local measure-ment	Loading Aeration: Decreased mixing
Vrábel, [Vrá99]	22000	4Rt	$5.5 \cdot 10^5$ - $11 \cdot 10^5$	80 - 600	0 - 0.5	Local measure-ment	Loading Aeration: No influence
Vrábel, [Vrá00]	8000	3Rt	$15 \cdot 10^5$	400 - 1500	0 - 2	Local measure-ment	Loading Aeration: No influence
Guillard, [Gui03]	1800-22000	2Rt to 4Rt	$3 \cdot 10^5$ - $14 \cdot 10^5$	80 - 2000	0 - 1	Local measure-ment	Loading and Flooding Aeration: Both effects

Due to the complexity of the influence of the gas flow on the flow pattern, only few correlations are available to describe the aerated mixing time. Three of them are presented in Table 2-6.

Vasconcelos [Vas95] and Alves [Alv95] proposed two different correlations (eq. (2.43) and eq. (2.44)) for aerated systems which are obtained in an acrylic glass reactor with a diameter of $D = 0.3$ m and for multiple impeller systems. Eq. (2.43) was determined in the loading regime and assumes that the influence of the aeration can be subtracted as a factor from the unaerated dimensionless mixing time ($t_{mix}^0 \cdot n$). This correlation gives the same trend for both, the unaerated and aerated system. Eq. (2.44) was established in the flooding regime and is an empirical correlation fitted to measurements. It is only a function of the factor Fl/Fl_c and three constants which are not described in more detail.

Table 2-6: Correlations of dimensionless aerated mixing time for multiple impeller and baffled stirred tank reactors

	Author	Formula		System
Two-phase	Vasconcelos, [Vas95]	$t_{mix}^g \cdot n \left(\dfrac{P_g}{P_0}\right) = t_{mix}^0 \cdot n$ $- 100\,[2^{n_s} - (n_s - n)]Fl$	(2.43)	$D = 0.3\text{m}$, $D = 0.2\text{m}$; Water/air $H/D < 3$
	Alves, [Alv95]	$t_{mix}^g \cdot n \left(\dfrac{P_g}{P_0}\right) = A + B\exp\left(-C\left(\dfrac{Fl}{Fl_q}\right) - 1\right)$	(2.44)	$D = 0.3\text{m}$, $D = 0.2\text{m}$; Water/air $H/D < 3$
	Shewale and Pandit, [She06]	$t_{mix}^g \cdot n =$ $453\,\left(w_g^0\right)^{0.127} + \left(\dfrac{H\,\varepsilon_g}{D}\right)^{-0.496}\left(\dfrac{P_g}{P_0}\right)^{-1.297} \cdot$ $\left(\dfrac{n^2 D}{g}\right)^{1.756} Ri_o^{0.711}$	(2.45)	$V_{fill}=0.07\text{m}^3$ water/air $H/D \geq 1$

2.5   Mass transfer in aerated stirred tank reactors

Especially for aerobic processes, the supply of oxygen is of high importance. A good product quality can only be achieved if the oxygen concentration can be set purposefully. For this reason, the liquid is aerated with air or pure oxygen. A transfer of mass from the gaseous phase with a high concentration of oxygen to the liquid with the lower concentration is taking place, resulting from the concentration difference Δc. The volumetric mass transfer coefficient $k_L a$ is the measure of the mass transfer performance and is one of the most important parameters for the design and the scale-up of stirred tank reactors. Thus, a lot of studies have been conducted in the last decades [Gar09], [Gar04], [Alv04], [Tat91], [Lin05a].

In the following chapter, the basic mass transfer mechanisms are explained. This is followed by a short overview of important mass transfer models and correlations.

2.5.1 Mass transfer in two phase systems

Mass transfer is a combination of two different mechanisms. Diffusion is caused by molecular motion, whereas convection is based on the movement of the surrounding fluid, which leads to a movement of entire areas ([Kra12], [Bae93], [Chr10]). The diffusion of a substance A into a system B is usually described by Fick's Law

$$\dot{n}_A = -D_{AB} \frac{dc_A}{dy} \tag{2.46}$$

with $\dot{n}_A$ the molar flow density, D_{AB} the diffusion coefficient and c_A the concentration of the substance A. Thus, the molar flow density is directly proportional to the concentration gradient $\frac{dc_A}{dy}$. Fick's second law describes the unsteady diffusion with

$$\frac{dc_A}{dt} = D_{AB} \frac{d^2 c}{dy^2} \tag{2.47}.$$

The second type of mass transfer is convection which is, beside of the material properties, strongly depending on the geometrical shape and the predominant flow. In contrast to diffusion, the substance is transported by the surrounding flow of the fluid. For a steady state convective flow, the molar flow density can be described with

$$\dot{n}_A = k_L (c_{A0} - c_{A\infty}) \tag{2.48}$$

with the mass transfer coefficient k_L and the concentration gradient $(c_{A0} - c_{A\infty})$ ([Bae93], [Chr10]).

Within a flow, typically both mechanisms, convection and diffusion, can be found. Due to the no-slip condition at the interface, the flow velocity is decreased in wall proximity and as a consequence, the convective mass transfer is decreasing whereas the diffusion part is increasing close to the interface. This leads to a formation of a concentration boundary layer analogous to the velocity boundary layer which can be seen schematically in Figure 2-1 ([Bae93], [Chr10]).

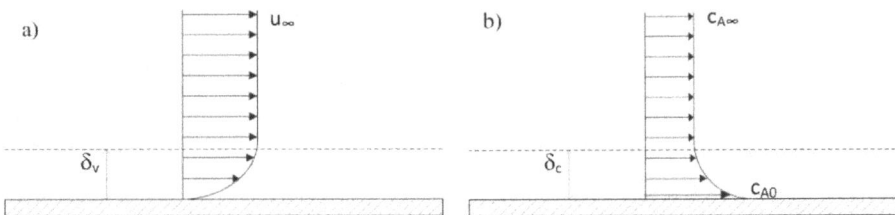

Figure 2-29: Schematic graph of the boundary layer - a) velocity boundary layer - b) concentration boundary layer [Inc02]

Assuming the no-slip condition at the interface the combination of equation (2.46) and (2.48) leads to

$$-D_{AB} \left(\frac{dc_A}{dy}\right)_{y=0} = k_L(c_{A0} - c_{A\infty}) \qquad (2.49)$$

and thus for the mass transfer coefficient

$$k_L = \frac{-D_{AB} \left(\frac{dc_A}{dy}\right)_{y=0}}{c_{A0} - c_{A\infty}} \qquad (2.50).$$

Different models and theories exist in literature to describe the mass transfer coefficient. A frequently proposed model is the two-film theory by Lewis and Whitman (1924), in which a simplified concentration curve at the phase boundary is used. The film theory assumes that all mass transfer occurs one-dimensionally within the thin boundary layer and that equilibrium is present at the interface. Due to Fick's second law (eq. (2.47)) a linear profile is forming for a stationary mass transfer and thus, with the integration of equation (2.49), leads to

$$-D_{AB} \frac{dc_A}{dy} = -D_{AB} \frac{c_{A\delta} - c_{A0}}{\delta} = \frac{D_{AB}}{\delta}(c_{A0} - c_{A\delta}) \qquad (2.51).$$

With equation (2.48), the mass transfer coefficient k_L

$$k_L = \frac{D_{AB}}{\delta} \qquad (2.52)$$

can be directly described with the diffusion coefficient and the thickness of the boundary layer. This interrelation shows that the mass transfer coefficient is dependent on the thickness δ of the boundary layer. The thinner the boundary layer, the larger the mass transfer coefficient gets ([Bae93], [Chr10], [Bra71]).

However, the theoretical assumptions of a laminar boundary film of thickness δ simplifies the process significantly compared to real gas-liquid systems, since both movements and turbulences in the phase boundary are excluded. Thus Higbie (1935) developed the penetration theory for unsteady mass transport. This is shown schematically in Figure 2-30, where a fluid element enters the phase boundary at time t_0 from the bulk flow. There it remains until time t_1, whereby diffusion of the gas component into the fluid element takes place during the mean residence time t_M.

Based on the Fick's second law (2.47), the integration under consideration of the boundary conditions

1. $c_{A,i} = c_{A,i,\infty}$ $t = t_0$ and $x \geq 0$

2. $c_{A,i} = c_{A,i}^*$ $t > 0$ and $x = 0$

3. $c_{A,i} = c_{A,i,\infty}$ $t > 0$ and $x \to \infty$

results in the molar flow density

$$\dot{n}_A = 2 \cdot \sqrt{\frac{D_{AB}}{\pi \cdot t_M}} \cdot (c_{A0} - c_{A\infty}) \qquad (2.53)$$

and leads to the mass transfer coefficient

$$k_L = 2 \cdot \sqrt{\frac{D_{AB}}{\pi \cdot t_M}} \cdot \qquad (2.54).$$

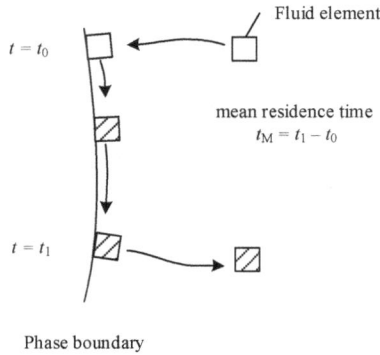

Figure 2-30: Visualisation of the mean residence time of a fluid element at the phase boundary in the penetration theory according to Higbie (1935)

Higbie calculated the mean residence time from the ratio of bubble diameter d_b to velocity u_b according to

$$t_M = \frac{d_b}{u_b} \qquad (2.55).$$

Thus, the mass transfer coefficient can be described as a function of the relative bubble velocity u_b and the dimensionless Reynolds and Schmidt numbers as

$$k_L = 1.13 \cdot u_b \cdot \sqrt{\frac{D_{AB}}{v_L} \cdot \frac{v_L}{d_b \cdot u_b}} = 1.13 \cdot u_b \sqrt{\frac{1}{Re\ Sc}} \qquad (2.56).$$

The determination of the mean residence time represents the main challenge in the application of the penetration theory. In addition, it is doubtful that an average residence time is sufficient to provide a reliable prediction of the mass transfer coefficient. For this reason, the renewal theory of Danckwerts, which is a modification of the penetration theory, was developed, that assumes different residence times at the phase boundary. Instead of the mean residence time t_M a residence time distribution $\varphi(t)$

$$\varphi(t) = s \cdot e^{-s \cdot t} \qquad (2.57)$$

with s the renewal factor, which is equivalent to the newly formed surface per time unit divided by the total surface, is used. This leads to the mass transfer coefficient

$$k_L = 1.33 \cdot \sqrt{D_{AB} \cdot s} \tag{2.58}.$$

In addition to these three theories, numerous further developments are available in literature to include other influences, such as the tangential velocity components, the degree of moving phase boundary or turbulence theories. Nevertheless, these additions certainly lead to an improvement of the prediction, but are basically only further fitting parameters that have to be determined for individual processes and will be further discussed later.

Besides of the mass transfer coefficient, the driving force, the concentration gradient, is an important factor in the overall mass transfer. In general, the gradient is defined by the saturation concentration c_A^*. Both boundary layers of the interphase, the one in the gaseous phase and the one in the liquid phase, need to be taken into account for the total mass transfer coefficient (see Figure 2-31 and Figure 2-32). In the film theory equilibrium, is assumed at the interface so that with Henry's law

$$c_{A,L}^* = He \cdot \frac{c_{A,g}^*}{p} \tag{2.59}$$

the concentration in the liquid phase $c_{A,L}^*$ can be described with the concentration in the gaseous phase $c_{A,g}^*$, the pressure p and the Henry coefficient He ([Kra12], [Bae93], [Bra71]). In a gaseous phase it can be assumed that the concentration difference between the concentration at the boundary layer $c_{A,g}^*$ and the concentration in the bulk phase $c_{A,g}$ is so small that the saturation concentration $c_{A,L}^*$ at the liquid side can directly be calculated with the concentration in the gas bulk phase

$$c_{A,L}^* \approx c_{A,L}^{eq} = He \cdot \frac{c_{A,g}}{p} \tag{2.60}$$

so that the mass flux

$$\dot{n}_A = k_L \left(c_{A,g} \cdot \frac{He}{p} - c_{A,L} \right) \tag{2.61}$$

can be calculated with the concentration in the gaseous phase. For simplification, the saturation concentration at the boundary $c_{A,g}^*$ will be used in the following [Kra12].

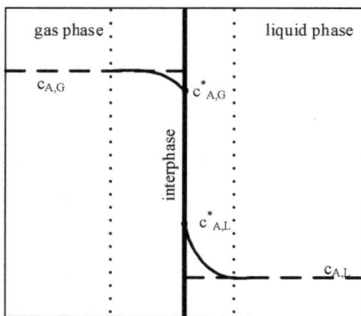

Figure 2-31: Correlation profile during mass transfer from a gaseous into a liquid phase

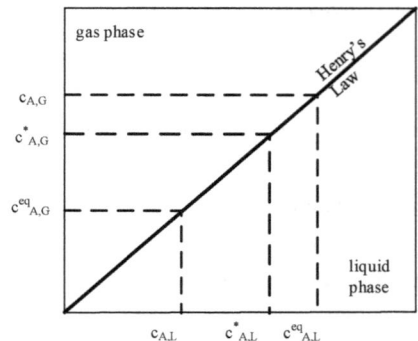

Figure 2-32: Function of concentration of substance A in a gas- and liquid phase at the interface

For the calculation of the total mass transfer $\dot{N}_A$, formula (2.48) needs to be multiplied with the total surface area A_{Gas}. Typically the specific surface area a ($a = A_{Gas}/V_{fill}$) is used instead so that the total mass transfer is

$$\dot{N}_A = k_L \cdot a \cdot V_{fill} (c_{A,L}^* - c_{A,L}) \tag{2.62}$$

with the volumetric mass transfer coefficient k_La and the filling volume of the reactor V_{fill}.

2.5.2 Mass transfer correlations

For the design of an apparatus, the processes involved in the flow behaviour must be precisely known. However, this is often not possible due to lack of literature data and insufficient experimental fundamentals. Therefore, the design is mainly achieved by means of integral measurements and empirical models. In order to ensure a scale transfer, the mass transport is often defined with the dimensionless Sherwood number Sh. Since the mass transport is decisively dependent on the velocity (Reynolds number) and the material system (Schmidt number, Sc), the Sherwood number is often described as a function of these dimensionless numbers in the form of

$$Sh = 2 + C_1 \cdot Re^a \cdot Sc^b \tag{2.63}.$$

For the coefficients, various values are available which are dependent on the process type and the process variable. Most investigations for this correlation were carried out in reactors without moving parts, such as a bubble column. An overview of the correlations for bubble columns is given in Table 2-7.

Table 2-7: Overview about Sherwood correlations developed in bubble columns

Author	Correlation	Info	
Brauer, [Bra81]	$Sh = 2 + 0.015\,Re^{0.89} \cdot Sc^{0.7}$	non-spherical bubbles	(2.64)
Hughmark, [Hug67]	$Sh = 2 + 0.0187 \left(Re^{0.484} Sc^{0.339} \left(\dfrac{d_b^3 g}{D_i^2}\right)^{0.024} \right)^{1.61}$		(2.65)
Reuß, [Reu70]	$Sh = 0.63 Re^{\frac{1}{2}} Sc^{\frac{1}{2}} \left(\dfrac{1 - \varepsilon_g}{1 - \varepsilon_g^{1/3}}\right)^{1/2}$	gas hold-up	(2.66)
Akita and Yoshida, [Aki73]	$Sh = 0.5\,Sc^{0.5} \left(\dfrac{g\,d_b^3}{v_L^2}\right)^{\frac{1}{2}} \left(\dfrac{g\rho d_b^2}{\sigma}\right)^{3/8}$		(2.67)
Montes, [Mon99]	$Sh = 1.33\,(Re \cdot Sc)^{0.5} \cdot (1.1 + 0.027 \cdot We^{0.5})$	oscillating bubbles	(2.68)

Correlations for stirred tank reactors are mainly presented in the dimensioned form of the k_L value. Most of the available correlations are empirical and some are based on theoretical assumtions [Gar04]. Two basical different approaches for the mass transfer coefficient k_L can be found in literature. Both theories are based on the widely accepted Higbie's penetration theory with a non-stationary diffusion in the gas-liquid interface and the contact time t_M.

The "slip velocity" model assumes that a mean gross flux of liquid relative to the bubble is present and the surface of the bubble renews itself. The surface renewal rate is only a function of the rise velocity of the bubbles. In the approach of Higby, the bubbles are divided by size, where the small bubbles ($d_b < 1$ mm) behave as a rigid spheres and the large bubbles ($d_b > 2.5$ mm) show a mobile interface. For the two correlations

$$k_L^{\text{rigid}} = 0.6 \sqrt{\frac{u_b}{d_b}} \cdot D_{AB}^{\frac{2}{3}} \cdot v^{-1/6} \tag{2.69}$$

$$k_L^{\text{mobile}} = 1.13 \sqrt{\frac{u_b \cdot D_{AB}}{d_b}} \tag{2.70}$$

only the rising velocity u_b, the diffusion coefficient D_{AB}, the diameter of the bubble d_b and the viscosity v need to be considered ([Lin05b], [Lin04]).

The second theory is the "eddy" model, initially presented by Kawase and Moo-Young (1988) [Mar08a], which assumes that a renewal of the liquid boundary layer takes place by small, vortex-shaped turbulences hitting the interface of the bubble ([Lin04], [Lin05a]). It has been shown that the mass transfer in stirred tanks strongly depends on the dissipated energy ([Lam70]) and that the surface renewal rate is higher than that for free rising bubbles. Consequently, the renewal rate needs to be a function of the eddies or the turbulence. The characteristic time scale can be calculated by the ratio of the two characteristic parameters of the Kolmogoroff's theory on isotropic turbulence, the eddy length η_K and the fluctuation velocity u_K according to

$$\eta_K = \left(\frac{v^3}{\varepsilon_T}\right)^{1/4} \tag{2.71}$$

$$u_K = (v \cdot \varepsilon_T)^{1/4} \tag{2.72}$$

with the specific dissipated energy ε_T in the fluid and the kinematic viscosity v leading to the penetration time

$$t_M = \left(\frac{v}{\varepsilon_T}\right)^{1/2} \tag{2.73}.$$

The combination of equation (2.54). and (2.73) leads to the mass transfer coefficient for "eddy" model

$$k_L = \frac{2}{\sqrt{\pi}} \cdot \sqrt{D_{AB}} \cdot \left(\frac{\varepsilon_T}{v}\right)^{1/4} \tag{2.74}.$$

Both theories are based on different assumptions concerning the influence of the energy input on the mass transfer coefficient. With the "eddy" model, the mass transfer coefficient k_L is increasing with increasing energy input due to a growing impact of the turbulence on the interface of the bubbles. The assumption of the "slip velocity" model is contrary, because an increased energy input usually leads to the formation of smaller bubbles with more stable surfaces and smaller rise velocities which then leads to a smaller mass transfer coefficient. Data can be found in literature, which supports both models ([Lin05b], [Lin04]).

In real processes, the interrelations are very complex and especially where the exact interphase is not known precisely, it is almost impossible to determine the mass transfer coefficient. Therefore, the mass transfer coefficient k_L and the specific surface area a are often combined to the volumetric mass transfer coefficient $k_L a$ that describes the efficiency of a process ([Alv04], [Lin05b], [Lin05a], [Lin04]). The volumetric mass transfer coefficient is an important parameter for stirred processes and is often used to compare different process parameters or scale-up models. An approved correlation for the calculation is

$$k_L a = C \cdot \left(\frac{P}{V}\right)^\alpha \cdot (w_g^0)^\beta \cdot \eta^\gamma \qquad (2.75)$$

with C as a geometrical parameter and α, β and γ as fitting coefficients. Most authors waive the viscosity in their correlation when water is used as liquid medium. Different values for the constants and exponents C, α and β are given by different authors for specific reactor volumes V_{fill}. Some examples can be found in Table 2-8.

Table 2-8: **Volumetric mass transfer correlation parameters for formula (2.36)**

Author	System	Agitator	Reactor volume V_{fill} / L	C $\cdot 10^{-3}$	α	β	γ
Van't Riet, [Van79]	Water with ions	Independent	2 - 2600	26	0.7	0.2	-
Smith et al., [SVM77]	Water	Disc	200-5000	10	0.475	0.4	-
Zhu et al., [ZBW01]	Water	Disc	45	31	0.4	0.5	-
Linek et al., [Lin04]	Water	Rushton	20	10	0.699	0.581	-
Gill et al., [GAB08]	Water with ions	Rushton	0.2	224	0.35	0.52	-
Vilaca et al., [VBF00]	Water with sulphite	Rushton	7	676	0.94	0.65	-

2.6　　　　Conclusion

From this literature survey it becomes clear that mixing times and mass transfer performances are among the most important parameters for aerated stirred tank reactors and that these reactors have been investigated intensively in the past. Large amounts of correlation for mixing times in single phase agitators for different reactor types are available in literature. Nevertheless, these correlations are mainly developed in small scale systems and large uncertainties arise when they are transferred to industrial scale systems. Additionally, only few investigations have been conducted on the interaction of the turbines with the gaseous phase, the resulting gas hold-up distribution and its influence on buoyancy driven flows and mixing.

Uncertainties also arise for the application of mass transfer correlations. Usually, the mass transfer performance is described with the volumetric mass transfer coefficient $k_L a$ as a function of power input and superficial gas velocity. Large amounts of correlations can be found in literature but none of them are achieving a general form so that each correlation is only suitable for a small range of parameters. The two phenomena, the mass transfer through the film k_L and the specific surface area a, are too complex to be described in one value. For a better description, the values need to be described and measured separately. However, especially in large scale systems, the information of the specific surface area is difficult to obtain due to a lack of optical access and thus only few investigations have been conducted into a separated measurement of mass transfer performance and specific surface area and the influence of the heterogeneity on these parameters.

In order to be able to model and to transfer processes from small scale to large scale, a profound understanding of the local phenomena such as the bubble size distribution, the local gas hold-up and the mixing time is crucial. For this reason, a transparent acrylic glass reactor on industrial scale has been designed that enables deep insights into aerated stirred processes by optical investigations.

In this thesis, the focus is placed on the influence of inhomogeneity of the gaseous phase on the overall mixing time and the mass transfer performance. Local investigations in a small scale 3 L reactor and an industrial scale of up to a 12 000 L reactor are made and compared to point out the important differences between these scales.

3. Experimental setup and measurement procedure

The experiments have been conducted in two different scales to investigate more closely the differences between lab scale and industrial scale. The lab scale reactor is a 3 L glass reactor, whereas the large scale is a 12 000 L acrylic glass reactor. In the following the facilities are presented, followed by a detailed description of the experiments.

3.1   Facilities

3.1.1   Laboratory scale – 3 L glass reactor

A glass reactor with a volume of 3 L is used for the laboratory-scale experiments. The reactor has a filling volume of $V_{fill} = 3$ L with a diameter of $D = 0.13$ m and a height of $H = 0.3$ m. The reactor consists of one piece with a rounded bottom. The stirrer shaft is fixed to the reactor lid in order to ensure optimal bearing. The used probes are also attached to the lid. The reactor is made of glass and is used in the standard setup without fixed baffles. In this case, only the probes ensure that no vortex is formed. To be able to compare the lab scale reactor with the large scale system, a ring with the possibility to adjust up to four baffles is installed. The agitator consists of a Rushton stirrer ($d = 62$ mm) and a pitched blade stirrer ($d = 62$ mm). The Rushton stirrer is mounted at the bottom ($h = 56$ mm) and the pitched blade stirrer at the top with a distance between the impellers of $s/d = 1$. The stirring device (ViscoPakt-Torque Measurement – X7, HiTec Zang GmbH, max torque $M_{max} = 7$ Ncm, resolution $M_{Res} = 0.003$ Ncm, reprodicibility $M_{Rep} = 0.05$ Ncm) is equipped with a precise torque measurement system to be able to determine the induced power input for different stirrer frequencies. For the gassing different spargers can be used. The standard sparger is the L-sparger with seven orifices with a diameter of $d_O = 1$ mm. The used devices can be seen in Figure 3-1 and Figure 3-2.

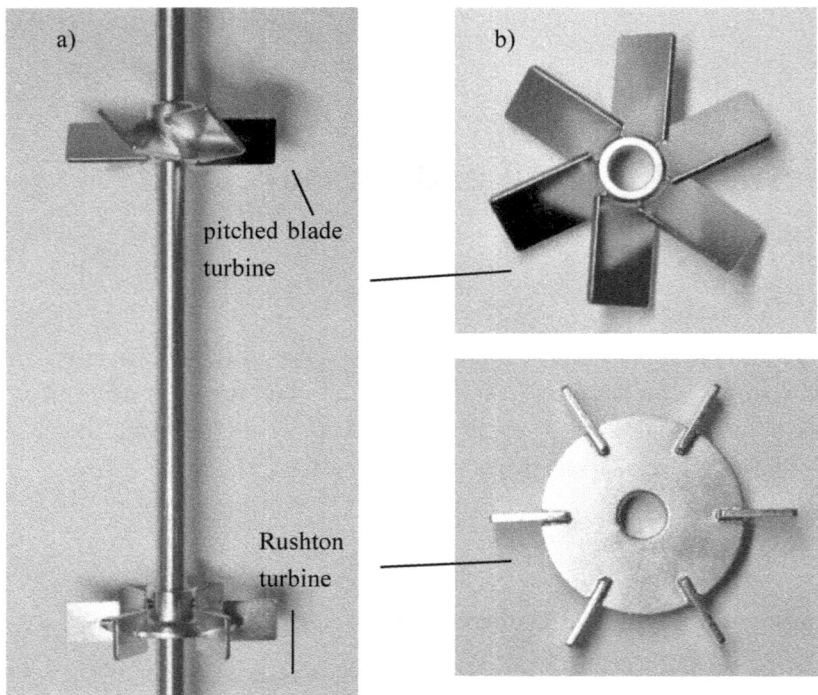

Figure 3-1: a) Standard stirrer combination (Rushton - pitched blade). b) Rushton turbine (bottom) and pitched blade turbine (top)

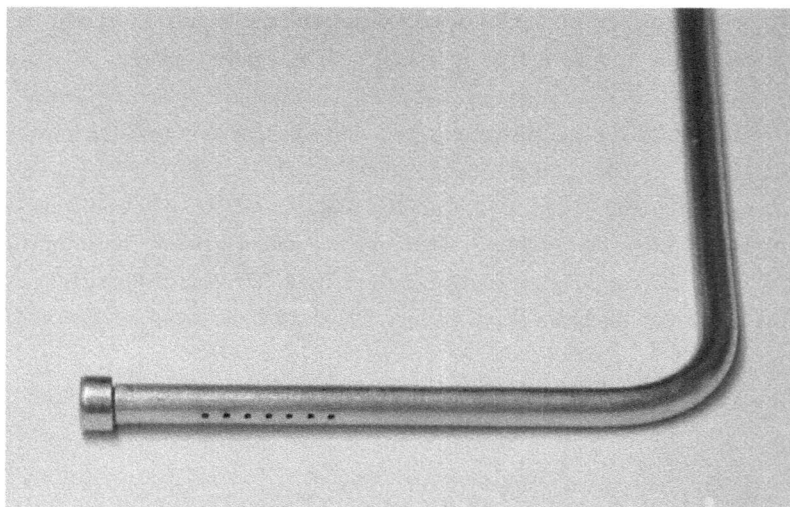

Figure 3-2: Standard L-sparger with seven orifices with a diameter of $d_O = 1$ mm

In Figure 3-3 the setup of the lab scale reactor can be seen. With a double jacket the temperature kept steady at $T = 37°C$. Different probes, such as pH, pO2 or pCO2, can be installed from the top lock. Nitrogen, oxygen or pressurised air can be used for aeration.

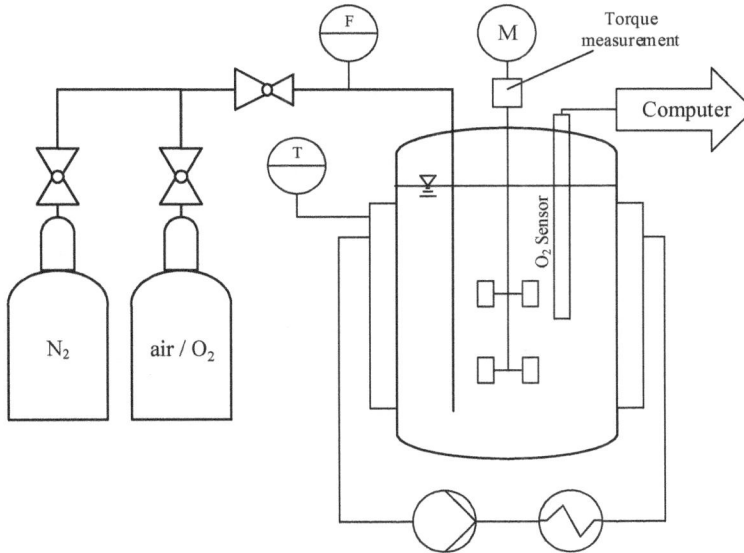

Figure 3-3: **Setup of the lab scale reactor**

3.1.2 Industrial scale – 12 kL acrylic glass reactor

The acrylic glass reactor, with a diameter of $D = 2$ m and a height of $H = 5$ m, has been erected at TUHH and is equivalent to the dimensions of the industrial reactors of Boehringer Ingelheim. For the experiments two different stirrer shafts can be used. The different stirrer shafts and stirrer combinations are shown in Table 3-1. The first one is the standard stirrer shaft which is used at the site in Biberach. With a mechanical seal the stirrer shaft is mounted to the motor, which is fixed under the reactor. The other three combinations are installed with the bottom mounted magnetic agitator (ZETA BMRF). It is possible to utilize different stirrer types at the shafts. In this work, mainly the combination of a Rushton turbine ($d = 0.665$ m at a height of $h = 0.61$ m) as the lower and a pitched blade ($d = 0.665$ m at a height of $h = 1.7$ m) (Rt-Pb) as the upper stirrer are investigated. Additionally, a second pitched blade stirrer can be mounted at the top of the stirrer shaft ($d = 0.665$ m at a height of $h = 0.15$ m) (Rt-2Pb).

Table 3-1: Investigated stirrer combinations

	Std-Rt-Pb	Rt-Pb-6kL	Rt-Pb-12kL	Rt-2Pb-15kL
V_{fill} /m³	12.5	6.5	12.5	15
H/D	2	1	2	2.7
Stirrer system	Upper: pitched blade Lower: Rushton turbine			stirrer pot Upper: pitched blade Middle: pitched blade Lower: Rushton turbine

Three baffles of width $L_B = D/10$ and a distance of 120° are mounted at the reactor wall. The temperature within the reactor is monitored by two temperature probes (Thermocoax Thermocouple TKI 20/10/NN, thermocouple type K, -200 to 1000°C, T90 according to EN 60584). With an external heat exchanger, the temperature level of the reactor can be kept constant. During the measurements, the external loop is switched off. Due to the lid and the good isolation characteristics of the acrylic glass, the temperature loss is 0.1°C/h and therefore an adiabatic state can be assumed. For the gassing an open tube sparger is installed at one of the baffles that can be adjusted from the outside with the orifice below the Rushton turbine (default setting), directly at the reactor wall or at any position in between. To measure the rate of gassing, a Coriflow (Proline Promass 80 F DN15, Endress+Hauser, accuracy ±0.35 %) is used. For the process gas, air, oxygen and nitrogen are available.

For the bottom mounted magnetic agitator, a torque measuring cell (Lorenz DR-3000) is installed at the stirrer shaft to precisely measure the induced torque by the impeller.

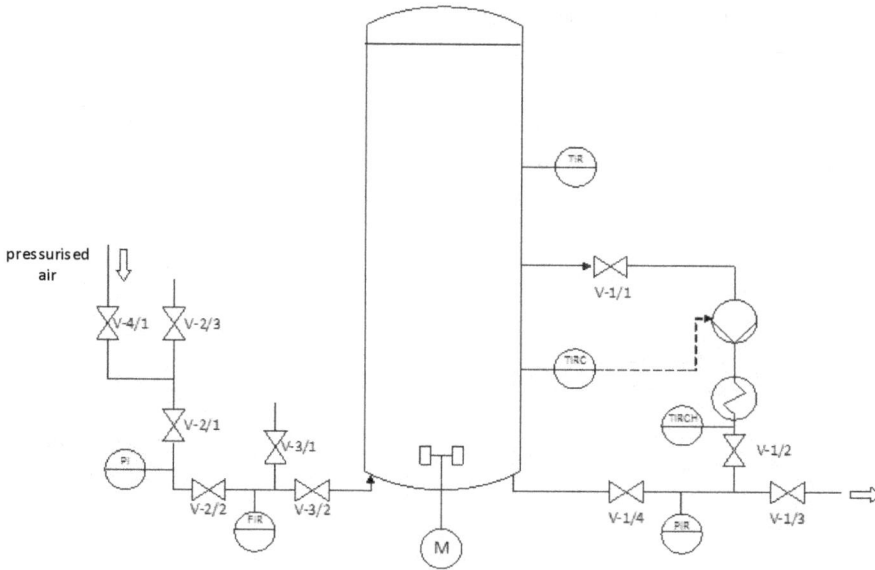

Figure 3-4: Flow chart of the plant of the 12 kL acrylic glass reactor

3.2 Determination of the energy input

The energy input is one of the major parameters for stirring processes and needs to be determined precisely. As described in chapter 2.1, with the torque measuring devices the torque M can be measured for a certain stirrer frequency n. With the formula $P = 2 \cdot \pi \cdot M \cdot n$ the induced power P and with the formula $Ne = \dfrac{P}{\rho \cdot n^3 \cdot d^5}$ the Newton number Ne can be calculated.

In the small scale system the torque was measured with the ViscoPakt-Torque Measurement system. Due to the small viscosity of water, the energy input to water is very small. Therefore, a minimum stirrer frequency of $n = 100$ rpm is needed to be able to measure the induced torque precisely without large errors. Below this value the idling torque gets too high and the measurement is not reliable anymore. The stirrer frequency of $n = 100$ rpm correspond to a Reynolds number of Re = 10 000. According to Zlokarnik [Zlo67a], a turbulent regime and thus a constant Newton number can be assumed. To be able to decrease the Reynolds number and receive reliable measurements in the laminar or transition regime, the viscosity of the used fluid needs to be increased. For the small scale system, this has been conducted by using glycerine and a glycerine water mixture at various temperatures. In Table 3-2 the used media with the corresponding viscosities and range of Reynolds numbers are presented. In the large scale system, only water can be used as the medium. This leads to a limitation of the investigated range of Reynolds numbers from $Re = 100\ 000$ to $Re = 2\ 000\ 000$ in the large scale system.

Table 3-2:   Used medium for the determination of the power input and the investigated range of Reynolds number

Reactor	Medium	Temperature	Density	Viscosity /m² s⁻¹	Range of Reynolds number
3 L-glass	Glycerine	37 °C	1250	$28.0 \cdot 10^{-5}$	$3 - 100$
		45 °C	1246	$16.2 \cdot 10^{-5}$	$10 - 200$
		50 °C	1242	$2.9 \cdot 10^{-5}$	$20 - 300$
	Glycerine-Water 2:1	45 °C	1170	$7.5 \cdot 10^{-6}$	$600 - 4\,000$
	Glycerine-Water 1:1	45 °C	1128	$3.1 \cdot 10^{-6}$	$2\,000 - 9\,000$
	DI Water	37 °C	993	$7 \cdot 10^{-7}$	$7\,000 - 40\,000$
12 kL-acrylic glass reactor	DI water	37 °C	993	$7 \cdot 10^{-7}$	$10^6 - 2 \cdot 10^7$

For a precise measurement of the induced torque, the idling torque that is not induced to the liquid due to its lost in the gear, needs to be known.

In the small scale system the idling torque can be determined by running the stirrer in air. In the large scale system, however, this is not possible because water is needed for greasing the bearing of the magnetic mounted agitator. When calculating the Newton number with the direct measured torque and neglecting the idling torque, the Newton number appears to be higher. Assuming both the idling torque and the Newton number are constant in the investigated range of stirrer frequencies, the idling torque can be determined iteratively until the calculated Newton number is constant. This is presented in Figure 3-5 and Figure 3-6. In Figure 3-5 the white symbols represent the direct measured torque. The corresponding calculated Newton numbers are plotted in Figure 3-6. It can be seen that for higher Reynolds numbers, the Newton number is constant. When the Reynolds number has decreased below the value of $Re = 5 \cdot 10^5$, the Newton number seems to increase. According to Zlokarnik [Zlo67a], the investigated flow regime should be in turbulent regime with constant Newton number. The increase of the Newton number can only be explained by the neglected idling torque. The filled symbols in Figure 3-5 represent the measured torque reduced by a constant idling torque. The new calculated Newton numbers are shown in Figure 3-6. It can be seen that a small reduction of the torque already leads to a large reduction of the calculated Newton number for small Reynolds numbers. Here, the Newton number is reduced from $Ne = 9$ to a value of $Ne = 6.5$. For higher Reynolds numbers, the reduction of the torque is almost without consequences. In this range the Newton number is reduced from $Ne = 6.6$ to a value of $Ne = 6.5$. Thus, it can be assumed that the iterative calculation of the idling torque gives reliable results for the Newton number.

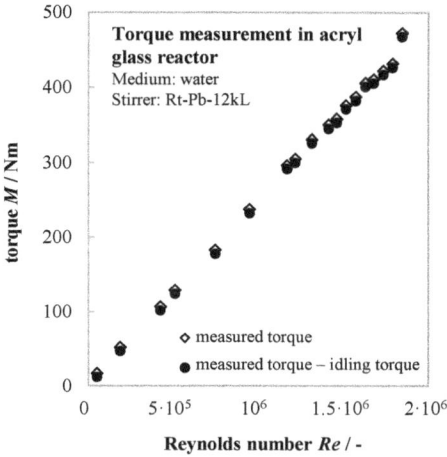

Figure 3-5:   Measured and calculated torque   Figure 3-6:   Calculated Newton number

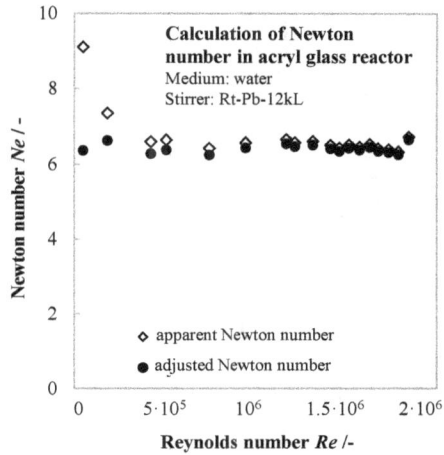

3.3 Determination of the bubble size distribution and gas hold-up

For the calculation of the specific surface area a, the bubble size distribution needs to be known precisely. There are different procedures available for the determination. A common method is the optical method, where pictures of the bubbly flow are taken by a camera. For low gas hold-ups, the images are typically taken from the outside of the reactor. For higher gas hold-ups, an endoscope which is located directly in the bubbly flow and connected to the camera, is needed. Since the gas hold-up for bio processes with mammalian cell culture are typically very small compared to applications in chemical industry, pictures, taken from the outside, are still applicable.

3.3.1 Measurement technique to determine the bubble size distribution in small and large scale system

In the 3 L glass reactor, the images are taken from one side of the reactor with a Nikon camera D50 with a 60 mm macro lens. The images have a resolution of 1600 x 1200 pixels. A bright illumination was achieved by use of an LED light source that was installed at the opposite site of the camera across the reactor. The small size and the small gas hold-up allow the evaluation of the total specific surface area of the whole reactor volume.

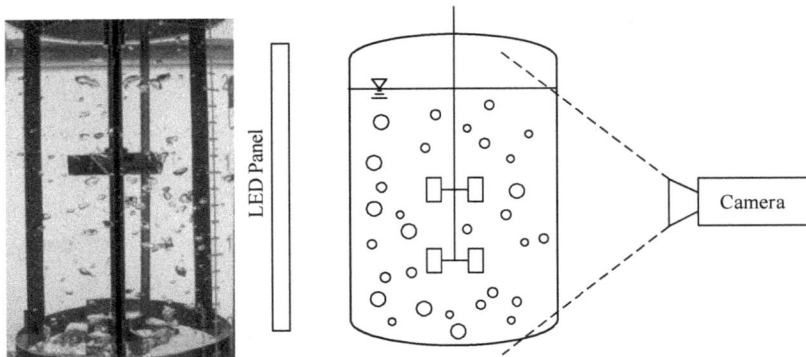

Figure 3-7:   Set up for bubble size measurement in the 3 L glass reactor

In the large scale reactor only a small volume can be evaluated due to the size of the reactor. The bubbly flow is evaluated above the stirrer shaft to ensure only a small optical distortion due to the arched surface of the reactor. In Figure 3-8 the position is illustrated and an exemplary picture is shown.

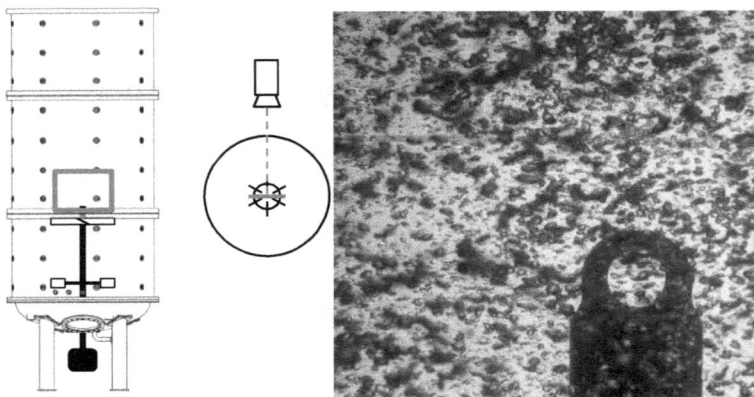

Figure 3-8:   (Left) Investigated area in the large scale system (over the stirrer shaft $H = 2$ m and $T = 1$ m) - (Right) exemplary picture of the bubbly flow

The pictures are taken with the same camera as in the 3 L glass reactor. For a better illumination, a flash light is installed at the back side of the reactor and a diffusor is attached at the reactor wall. Furthermore, a high shutter speed is chosen to prevent motion blur on the photographs. To eliminate the distortions, several photographs of a target are taken, using the exact same settings for the camera. The target consists of a piece of laminated graph paper and is placed in the middle of the reactor. This way, the distortions can be eliminated and furthermore it enables the conversion of pixels into millimetres.

3.3.2 Evaluation of the bubble size distribution

The evaluation was done semi-automatically by the image processing program ImageJ. The evaluation procedure starts with a binarisation of the pictures followed by the marking of all bubbles that are located within the focused plane (see Figure 3-9 and Figure 3-10).

The shap of the bubble is strongly dependent on the size and on the surrounding liquid flow. Especially for larger bubbles, the shape can be very irregular. In order to determine the volume and the surface, these bubbles are assumed to be a rotational ellipsoid as can be seen in Figure 3-10. The volume of the ellipsoid is given by

$$V_b = \frac{4}{3} \cdot \pi \cdot a^2 \cdot b \tag{3.1}$$

with a as the major and b as the minor half-axis. The surface A of an ellipsoid is calculated according to

$$A_{\text{Ellipse}} = 2\pi a^2 \cdot \left(a + \frac{b^2}{\sqrt{a^2 - b^2}} \arcsin\left(\frac{\sqrt{a^2 - b^2}}{b} \right) \right) \tag{3.2}.$$

Figure 3-9:   Exemplary photograph of a bubble swarm in the 12 kL reactor

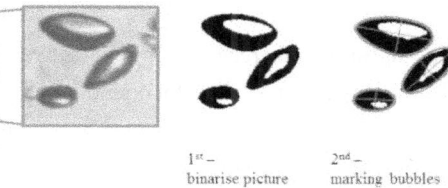

Figure 3-10:   Example of the evaluation procedure

The summation of all single volumes and surfaces results in the total volume $V_{g,tot}$ and the total surface $A_{g,tot}$, which can be used to calculate the Sauter mean diameter d_{32} according to

$$d_{32} = 6 \frac{V_{g,tot}}{A_{g,tot}} \tag{3.3}.$$

For each operation condition a minimum amount of 200 bubbles are evaluated. With the evaluation of at least 200 bubbles the cumulative mean value stays within an error of $\pm$ 5 % as can be seen in

Figure 3-11 for the 3 L reactor at a power input of $(P/V) = 70$ W/m³ and a superficial gas velocity of $w_G^0 = 0.63$ mm/s.

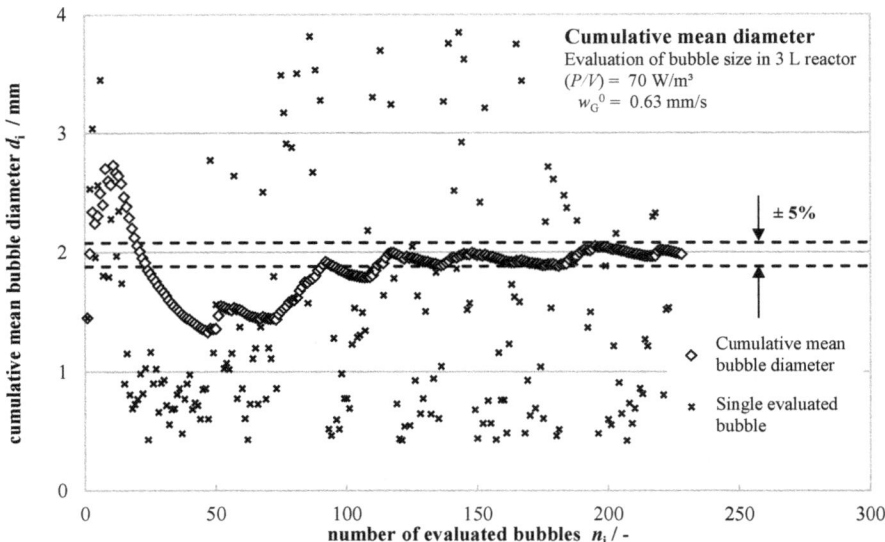

Figure 3-11:   Cumulative mean value of the equivalent bubble diameter for one operation point

3.3.3   Evaluation of the gas hold-up

For the gas hold-up in the 3 L reactor, all bubbles are evaluated in order to determine the total gas volume V_G present in the system. The total amount of gas in the system $V_{g,tot}$ divided by the reactor volume V_{fill} gives the gas hold-up

$$\varepsilon_g = \frac{V_{g,tot}}{V_{g,tot} + V_{fill}} \tag{3.4}.$$

In the large scale system the gas hold-up ε_g is estimated by measuring the change in the liquid height under aerated H^g and unaerated H^0 conditions according to

$$\varepsilon_g = \frac{H^g - H^0}{H^g} \tag{3.5}.$$

3.4   Determination of the mixing time

To determine the mixing time, different methods are possible depending on the design of the reactor and the type of medium. The most simple and common method is the determination by logging the conductivity. Thereby a saline solution is added to the reactor at the time t_0 while simultaneously the conductivity is measured at different positions within the reactor. By adding the saline solution locally, the conductivity rises with different velocities at the different probe positions depending on the operating conditions. When the conductivity reaches 95 % of the end value at all probes, the mixing is complete and the mixing time t_{95} can be calculated. The advantage of this method is the simple implementation as well as the possibility to conduct the method in industrial reactors without optical access. However, only selective assumptions can be made since the conductivity can only be measured at defined positions and usually only close to the reactor walls. Possible dead zones e.g. behind baffles or within the reactor might remain undetected.

For a reactor with optical access another option is the method of decolouration. For this, 200 mL of a 5.8 vol-% phenolphthalein solution is added to the reactor once. Phenolphthalein is colourless for pH values between 0-8.2 and is coloured pink in case of pH values above 8.2. At the beginning of each measurement, the pH-value is shifted to a higher value by adding 500 mL of a 2 molar NaOH solution, which leads to a change in the colour of the phenolphthalein ion. After adding HCl in a 1.5 times higher stoichiometric ratio (N_{HCl}/N_{NaOH} = 1.5), the pH-value is shifted backwards and the phenolphthalein turns colourless. The mixing time is estimated by measuring the time that is needed from the addition of the HCl solution to reach a decolouration level of 95 %. The adding of the hydrochloride acid occurs at a position that is comparable to the feed position in the industrial reactor. Thereby the acid is added to the wall of the reactor and drops to the surface 5° behind a baffle in relation to the flow direction as can be seen in Figure 3-12.

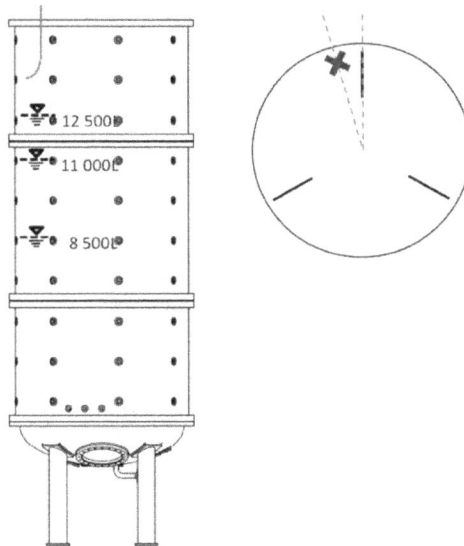

Figure 3-12:   Position of acid addition

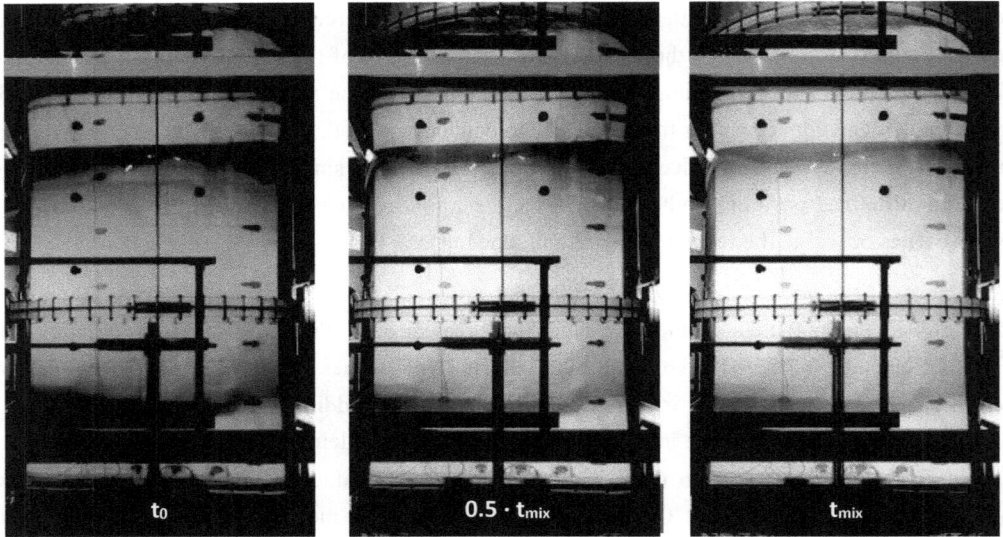

Figure 3-13: **Example of a decolouration time series**

Five spotlights and a diffusor at the backside of the reactor are used for a homogeneous illumination. For a quantitative evaluation of the mixing time, the decolouration is recorded by a CMOS camera with a sampling rate of 1 s^{-1}. The images are saved with a resolution of 4000x6000 pixels in the native raw format. Afterwards, the time series are analysed by using the image analysing software ImageJ that evaluates the development of the mean grey value over time. The plotted mean grey value for each time step is additionally normalized to the end value of the corresponding measurement (see Figure 3-14). The mixing time is defined as the time from the addition of the acid until the normalised grey value reaches 0.95.

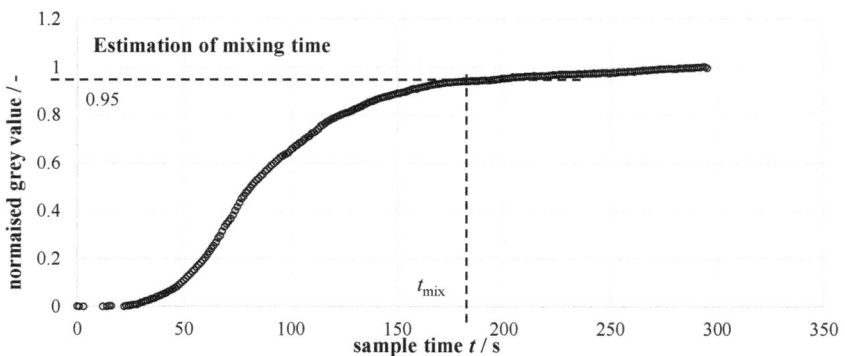

Figure 3-14: **Example of the normalised grey value over the sample time to determine the mixing time**

For further validation of the optical measurements, experiments have been repeated for 38 times at one operation condition. In Figure 3-15 the results of each measurement (cross) as well as the cumulative mean value (square) are shown. The single measurements fluctuate strongly around the mean value. Nevertheless it can be seen that after nine measurements the mean value reaches a stable terminal value with a deviation of less than 5 %. However, a slight drift with a rising numbers of experiments can be seen which can be explained by the salinisation due to the addition of sodium hydroxide and hydrochloride acid. For this reason, the number of experiments needs to be as low as possible to reduce the influence of the salinisation but also as high as necessary to ensure statistical certainty of the results due to the high fluctuations. The experiments in this work were repeated five times. With this number of experiments the variation is about 6 %. The change due to salinisation is 2 %.

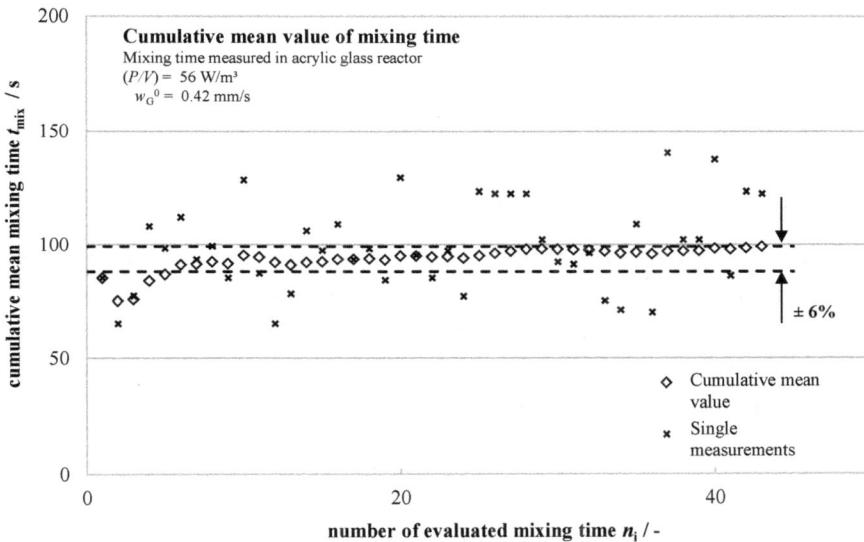

Figure 3-15:   Cumulative mean value of mixing time for one operation point

3.5   Measurement of the volumetric oxygen mass transfer

3.5.1   Oxygen probes (location, devices and calibration)

For the measurement of the oxygen concentration, a WTW galvanic dissolved oxygen sensor (FDO 925; measurement range 0 to 200 % DO saturation, limit of detection <0.1 % saturation value; response time (t90) <30 sec) was used. In the course of experiments, different positions of the probes have been tested to determine the influence of the position and calibration procedure on the oxygen level and measured mass transfer coefficient. The main results were obtained using the WTW-probe at the postions seen in Figure 3-16.

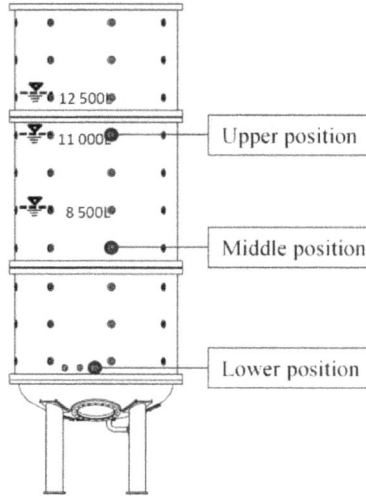

Figure 3-16: Position of oxygen probes

3.5.2 Measurement of mass transfer performance

For the determination of the mass transfer coefficient k_La, the dynamic method is used. For this oxygen is stripped out of the medium with nitrogen to an atmospheric oxygen saturation level below 15 %. Subsequently, the desired stirrer frequency n and the desired rate of gassing q are set and the increase of the concentration of oxygen c_{O2} is being recorded. The plot of the recorded oxygen levels against the corresponding time t can be described according to

$$\frac{dc_{O_2}}{dt} = k_L a \left(c^* - c_{O_2} \right) \tag{3.6}$$

with the volumetric mass transfer coefficient k_La and the saturation concentration $c_{O_2}^*$. The equation can be transformed into a logarithmic expression

$$\frac{\ln\left(c_{O_2}^* - c_{O_2}'\right)}{\ln\left(c_{O_2}^* - c_{O_2}''\right)} = k_L a \left(t'' - t' \right) \tag{3.7}$$

with the concentration c_{O_2}' at the time t' and the concentration c_{O_2}'' at the time t''. Equation (3.7) provides the value of k_La as the slope of the logarithmic function of the oxygen plot. An exemplary evaluation can be found in Figure 3-17. The evaluation of the k_La value was done from a saturation of $c_{O_2} = 60$ % to $c_{O_2} = 80$ %.

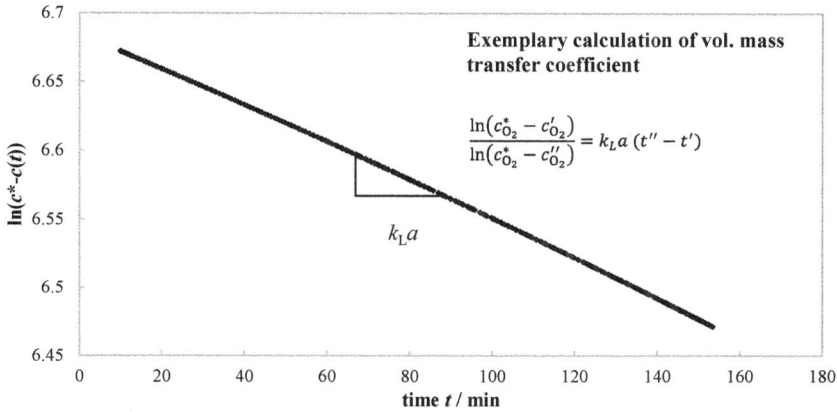

Figure 3-17: Exemplary evaluation of the volumetric mass transfer coefficient

Assumption for the calculation of the saturation concentration $c_{O_2}^*$

The saturation concentration $c_{O_2}^*$ is linked to the Henry coefficient and thus, it is a function of the partial pressure. In the small scale system it can be assumed that the partial pressure of oxygen in air for the whole reactor is $p_{O_2} = 0.21$ bar. Thus, the saturation concentration is $c_{O_2}^* \approx 7.3$ mg/L at a temperature of $T = 37°C$.

However, in the large scale system this assumption cannot be made. Due to the hydrodynamic pressure, the partial oxygen pressure is linearly increasing with increasing depth. With a filling volume of $V_{fill} = 12$ m³, the pressure at the bottom of the reactor is about 0.4 bar higher than at the top. As a consequence, a constant saturation concentration gradient over the liquid height appears, when aeration is on. By taking the partial pressure into account, a linear increase of the saturation concentration from $c_{O_2 \, BOTTOM}^* \approx 10.11$ mg/L at the bottom to $c_{O_2 \, TOP}^* \approx 7.3$ mg/L at the surface can be assumed (see Figure 3-18). Nevertheless, the stirring and the buoyancy driven flow leads to such a good mixing that the maximum oxygen concentration in the liquid phase that can be achieved by aeration is the average between $c_{O_2 \, BOTTOM}^*$ and $c_{O_2 \, TOP}^*$. This average concentration of $\overline{c_{O_2}^*} \approx 8.7$ mg/L can be measured at every point in the reactor and thus, it can be assumed to be the global saturation concentration for the calculation of the mass transfer coefficient (see Figure 3-19).

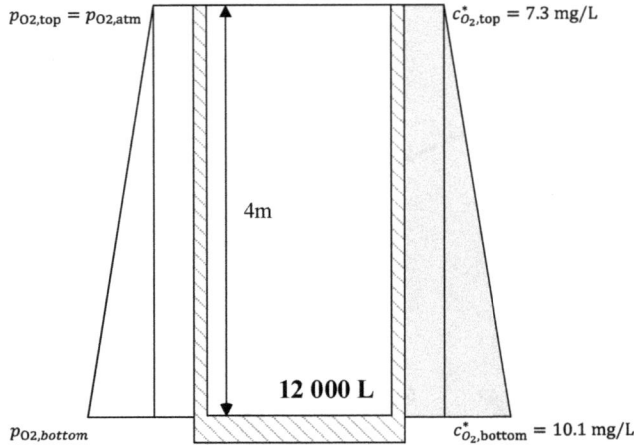

Figure 3-18: Partial pressure and theoretical saturation concentration distribution over the reactor height

In the figure: $p_{O2,top} = p_{O2,atm}$; $c^*_{O_2,top} = 7.3$ mg/L; 4m; 12 000 L; $p_{O2,bottom}$; $c^*_{O_2,bottom} = 10.1$ mg/L

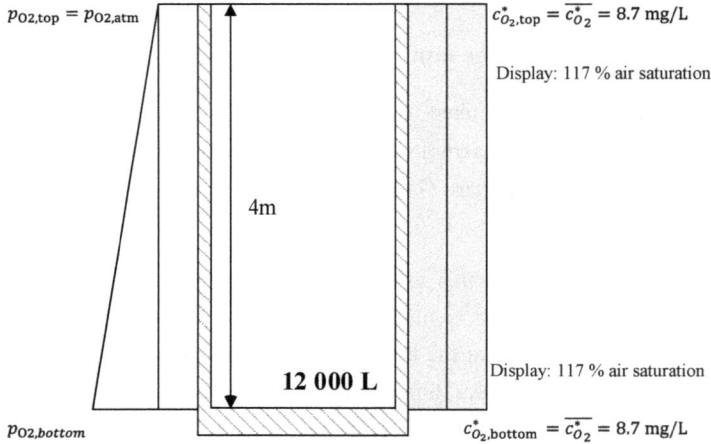

Figure 3-19: Measured distribution of oxygen concentration over the liquid height

In the figure: $p_{O2,top} = p_{O2,atm}$; $c^*_{O_2,top} = \overline{c^*_{O_2}} = 8.7$ mg/L; Display: 117 % air saturation; 4m; 12 000 L; Display: 117 % air saturation; $p_{O2,bottom}$; $c^*_{O_2,bottom} = \overline{c^*_{O_2}} = 8.7$ mg/L

This assumption can also be confirmed by looking at the oxygen concentration during aeration. In Figure 3-20 the trend of the oxygen concentration is plotted for the lower and for the upper probe as a function of the aeration time. A small shift between the probes can be seen at the beginning of the aeration. The final concentration values, however, are at both position of $c^*_{O_2} = 117$ % which corresponds to $c^*_{O_2} \approx 8.7$ mg/L. Thus, for all mass transfer evaluations, a saturation concentration of $\overline{c^*_{O_2}} \approx 8.7$ mg/L is used.

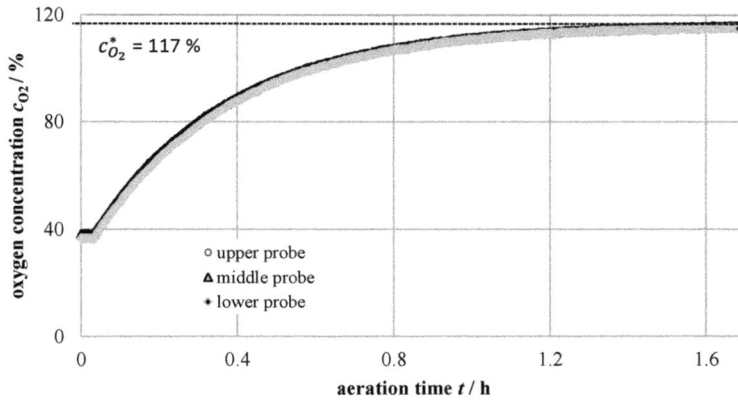

Figure 3-20: Increase of the oxygen concentration during aeration with air, measured at two positions

4. Results and discussion

4.1   Energy input

The power input plays a particularly important role in mammalian cell cultivation, since a high energy input can be harmful for the cells and thus negative for the production. An easy way for the calculation of the power input without measuring it directly is using the dimensionless Newton number Ne. It has been shown that the Newton number stays constant at different scales as long as geometrical ratios such as the level ratio (H/D) or the ratio of the stirrer to tank diameter (d/D) remain constant. However, if different stirrers are combined, or if size ratios are changed, the Newton number must be determined again. This also applies to the presence of a gaseous phase, since the influence on the power input cannot be predicted precisely yet. In the following, chapter initially the Newton numbers for the used stirrer combinations are determined in the lab scale and the industrial scale reactor. Afterwards, the influence of the gaseous phase on the power input in the two scales is investigated and compared.

4.1.1   Energy input in single phase flow

Small scale system

To measure the induced torque for a wide range of Reynolds numbers, the viscosity needs to be varied. This has been done in the small scale system by using different combinations of water and glycerine at various temperatures. This procedure extends the measurable range of Reynolds number, leading to an investigated range from $Re = 2$ to $Re = 9 \cdot 10^4$.

To verify the plausibility of the results, a known system was measured to compare the results with literature data. For this purpose, a simple Rushton turbine in a system with four baffles is used. In Figure 4-1, the measured Newton number for a single Rushton stirrer for the different mixtures can be seen. The error of each data point is calculated over the fluctuation in the stirrer frequency during the measurement. It can be seen that in the range of $Re < 100$ the Newton number is decreasing with increasing Reynolds number. In the range of $Re = 10^2$ to $Re = 10^3$ it comes to an undershoot of the Newton number. For higher Reynolds numbers the Newton number becomes constant with a value of $Ne = 4$.

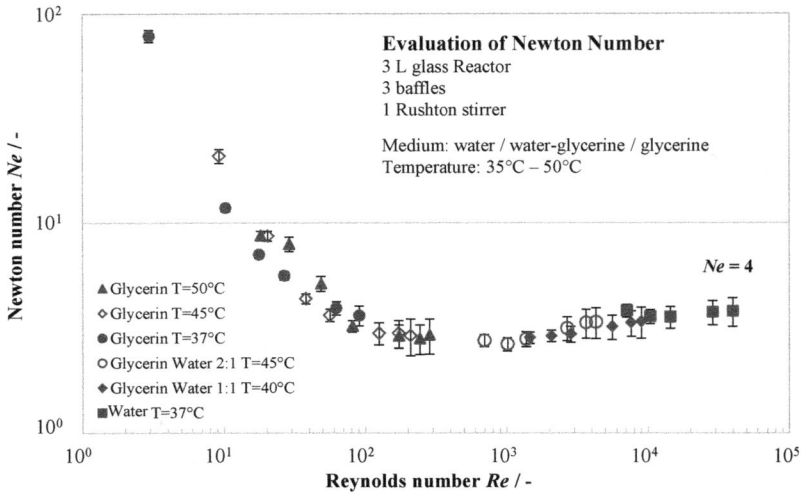

Figure 4-1:   Newton number over Reynolds number in the baffled 3 L glass reactor for one Rushton turbine

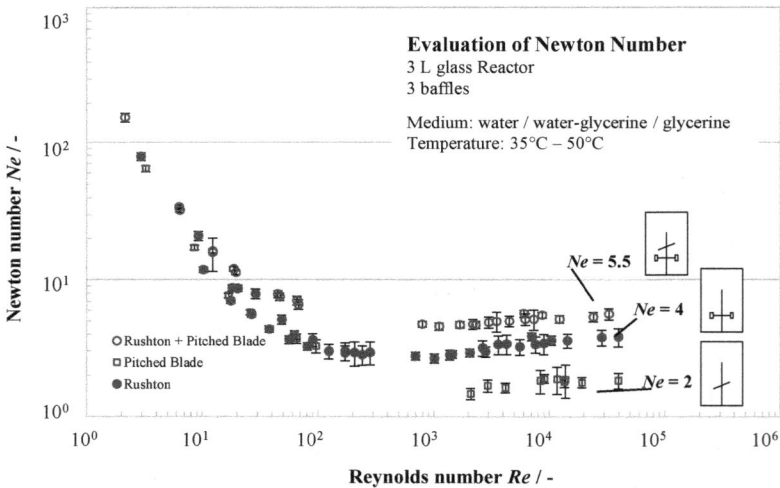

Figure 4-2:   Newton number over Reynolds number in 3 L glass reactor for Rushton turbine, pitched blade and the combination of both in the baffled 3 L glass reactor

In addition to the single Rushton turbine a single pitched blade and a combination are also tested. The results are shown in Figure 4-2. As expected, the constant Newton number of the pitched blade is smaller than the Rushton turbine and reaches a value of $Ne = 2$. The combination of the Rushton and the pitched blade reaches a value of $Ne = 5.5$. It becomes clear that with a combination of stirrers the final Newton number does not necessary need to be the sum of the individual Newton numbers.

In Figure 4-3 the results are compared with the literature data of Zlokarnik [Zlo67a]. It can be seen that the results for the Rushton turbine are in good agreement to the literature data. However, it has to be noted that the geometrical proportions do not accord with each other which is why the results are not totally comparable. For the pitched blade stirrer, a Newton number of $Ne = 2$ was measured. This value lies between the classical axially pumping stirrers, such as the impeller (ib) with $Ne = 0.75$ or the propeller (hb) with $Ne = 0.35$, and the radially pumping turbine (gb).

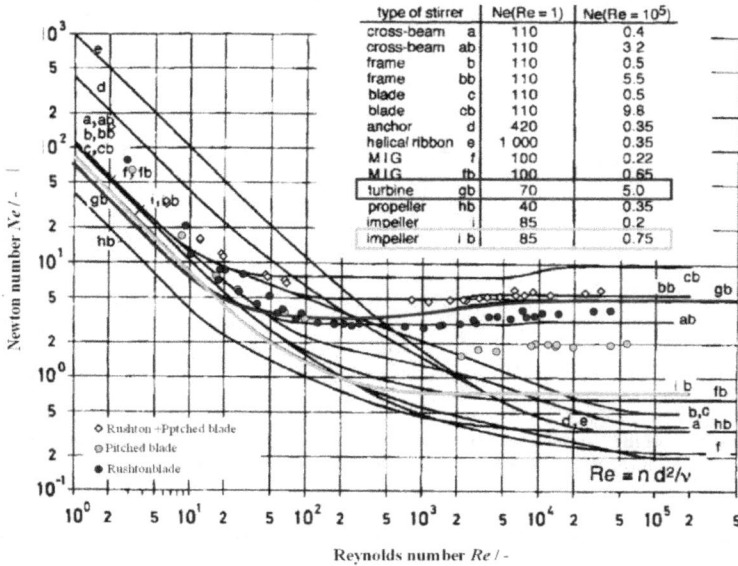

type of stirrer		Ne(Re = 1)	Ne(Re = 10^5)
cross-beam	a	110	0.4
cross-beam	ab	110	3.2
frame	b	110	0.5
frame	bb	110	5.5
blade	c	110	0.5
blade	cb	110	9.8
anchor	d	420	0.35
helical ribbon	e	1 000	0.35
MIG	f	100	0.22
MIG	fb	100	0.65
turbine	gb	70	5.0
propeller	hb	40	0.35
impeller	i	85	0.2
impeller	i b	85	0.75

Figure 4-3: Comparison of the Newton number with literature data of Zlokarnik [Zlo03])

Large scale system

In the large scale system, only water can be used. This leads to a limitation of the measurable range of Reynolds numbers from $10^5 < Re < 2 \cdot 10^6$. To investigate the influence of the liquid heights on the Newton number, the experiments were carried out at different filling volumes. The results are presented in Figure 4-4 for the standard stirrer with the simple bearing design and in Figure 4-5 for the magnetic mounted agitator.

In the system with the simple bearing design it seems that the Newton numbers at lower Reynolds numbers are higher than the end value. However, this is only the case for the filling volumes of $V_{fill} = 8\,500$ L and $V_{fill} = 12\,000$ L. The value for the filling volume of $V_{fill} = 11\,000$ L is in the same range as the end value. In this system, an in-house measuring cell was used, which leads to high deviations at low impeller frequencies, so that it can be assumed that the difference in the Newton number is attributed to the measurement inaccuracy. Thus, in the examined regime the Newton number for the simple bearing system is constant at $Ne = 5.5$ for all filling volumes. This in turn is identical to the values found in the small scale reactor for the same impeller combination. This leads to the conclusion that for single phase conditions the Newton number can be easily determined in a small scale system with similar geometrical relation even across many orders of magnitude.

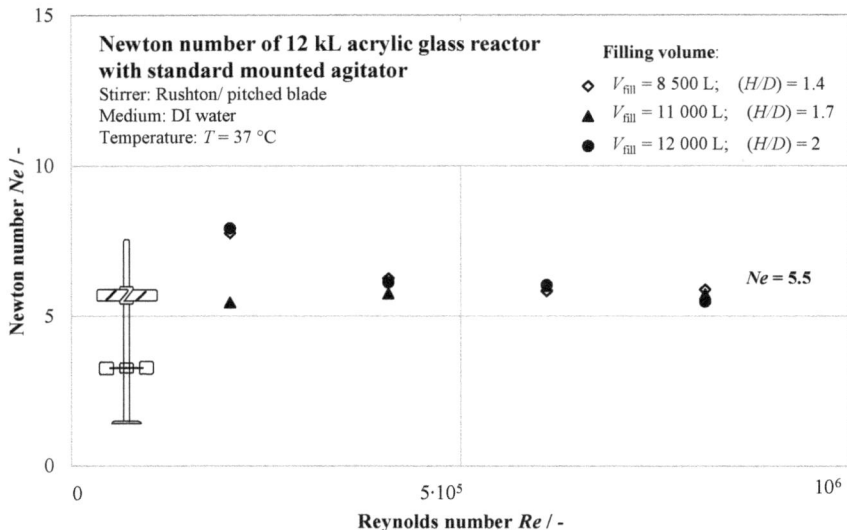

Figure 4-4: Dependency of the Newton number from the Reynolds number and the filling volume for unaerated condition in the large scale system with standard mounted agitator

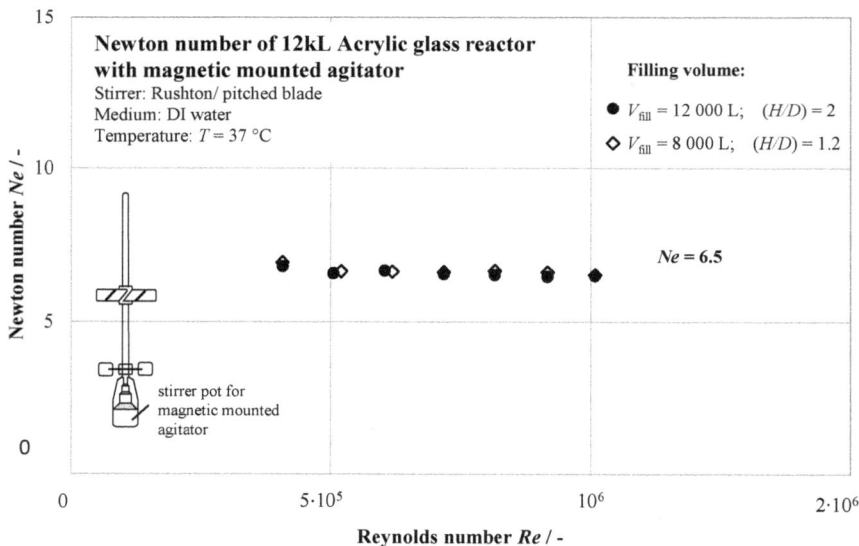

Figure 4-5: Dependency of the Newton number from the Reynolds number and the filling volume for unaerated condition in the large scale system with magnetic mounted agitator

In the system with the magnetic mounted agitator, it is possible to install a precise measuring cell, which delivers accurate results even at low stirrer frequencies. This can be seen in the results of the Newton number in Figure 4-5. In the whole range of the Reynolds number, a constant Newton number is found. Due to the stirrer pot, a slightly higher Newton number of $Ne = 6.5$ is reached for the same stirrer combination. However, it was found that the influence on mixing as well as on the mass transfer can be neglected compared to the influence of the Rushton and the pitched blade stirrer.

4.1.2   Energy input in two phase flow

Small scale system

In Figure 4-6, the influence of the superficial gas velocity on the power input is presented. For a constant superficial gas velocity and increasing Reynolds number, it can be seen that the Newton number is decreasing until a minimum Ne_{min} is reached. A further increase in the Reynolds number causes the Newton number to increase again. This trend is apparent for all superficial gas velocities. However, the value of the minimum Newton number as well as the corresponding Reynolds number are dependent on the superficial gas velocity. At $w_G^0 = 0.16$ mm/s the Newton number is only reduced to $Ne = 3.7$, which is 20 % less compared to the unaerated value of $Ne = 4.6$. At $w_G^0 = 0.79$ mm/s the measured minimum is $Ne = 2.4$. This means that the power input in the presence of a gaseous phase has decreased by 48 %. However, the decrease of the Newton number with increased Reynolds number seems to be independent of the superficial gas velocity. All values are located on the same line (Ne_{min}) until the corresponding minimum is reached.

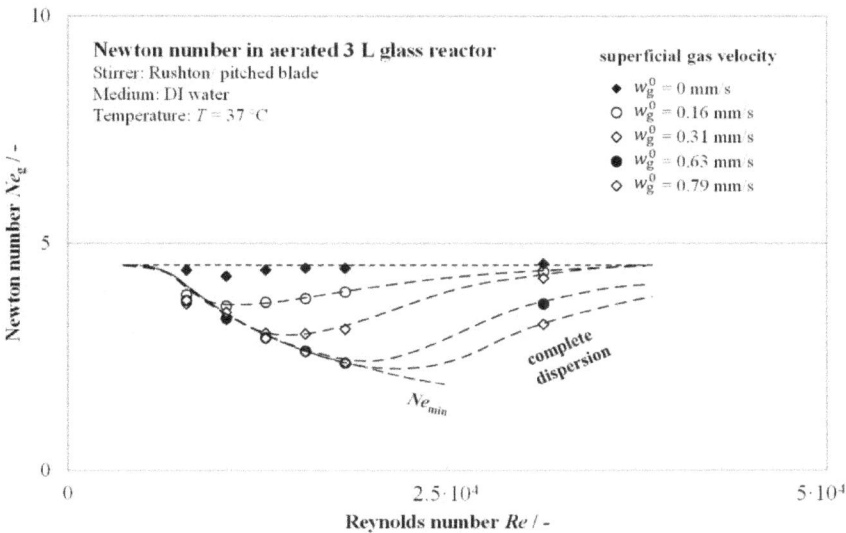

Figure 4-6:   Dependency of the Newton number on the Reynolds number at constant superficial gas velocities in the 3 L reactor

This behaviour is related to the cavities and thus to the loading and flooding regime in the reactor. For lower stirrer frequencies, the gas is passing the stirrer blades. With increasing stirrer frequency, the cavities behind the blades are increasing and thus more gas can be captured. These gas cavities are leading to a better hydrodynamic flow over the blades which leads to a decrease of radial liquid velocities and thus to a decrease in power input. At a certain point, the power input reaches a minimum. When the stirrer frequency exceeds this minimum, the power input starts to increase again. At this point, the stirrer frequency is high enough to disperse the gaseous phase and to increase the acceleration of the fluid. This in turn leads to an increase of the power input. Thus, for a higher superficial gas velocity, a higher Reynolds number is needed to start the dispersion of the gaseous phase.

This trend of the aerated Newton number is in agreement with the data from Kapic and Heindel [Kap06] which has been presented in chapter 2.3.3. Rearranging the data from Figure 2-25 leads to the graph in Figure 4-7, where the Newton number under aerated condition is plotted against the Reynolds number for different constant superficial gas velocities. This plot confirms that different minima of the Newton number exist for different gassing rates. Kapic and Heindel stated that these minima indicate the transition from loading regime to complete dispersion regime and are therefore owed to a change in the cavities.

For each minimum of the Newton number for both figures (Figure 4-6 and Figure 4-7), a Froude number Fr and a specific Flow number Fl_{CD} can be determined. Each Flow number can thus be assigned to a minimum Froude number, i.e. a minimum speed, which is required for a sufficient dispersion of the gaseous phase. The dependency of the critical Flow number from the Froude number is plotted in Figure 4-8.

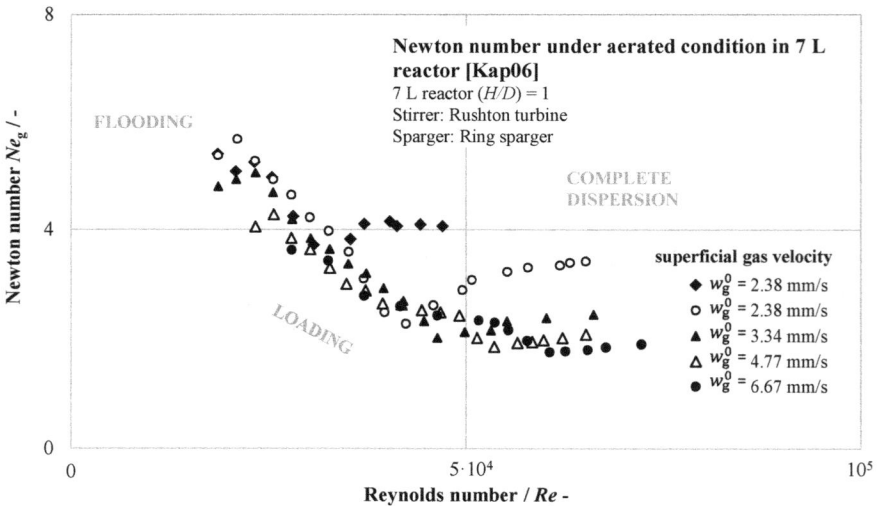

Figure 4-7: **Dependency of the Newton number on the Reynolds number at constant superficial gas velocities for a 7 L Reactor presented by Kapic and Heindel [Kap06]**

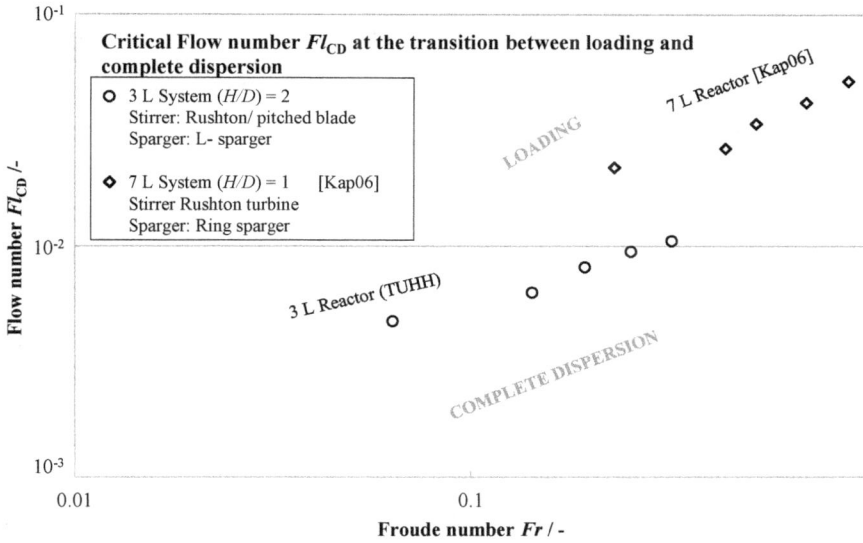

Figure 4-8: Critical Flow number between loading and complete dispersion regime in the 3 L reactor in comparison with results from a 7 L reactor from literature [Kap06]

For both systems it can be seen that with increasing Froude number the critical Flow numbe, from which a good dispersion can no longer be assumed, also increases. Furthermore, it becomes evident that in the 7 L reactor a slightly higher Flow number can be set until the end of complete dispersion is reached. This difference can be explained with the smaller level ratio of $(H/D) = 1$ in the 7 L reactor. For the same Froude number, a much larger power input can be obtained and thus the gaseous phase can be dispersed more efficiently. Nevertheless, despite their different designs, the results in both reactors are in good agreement.

Large scale system

The influence of the gassing rate on the Newton number in the large scale system is shown in Figure 4-9 as a function of the Reynolds number for different superficial gas velocities. In contrast to the small scale reactor, no significant change of the Newton number can be measured even at relatively high superficial gas velocities. In the range of small Reynolds number of $Re < 5 \cdot 10^5$, an increase in the Newton number for the aerated condition can be observed. So far it is not clear whether there is actually an increase in the Newton number, or if this is only caused by the measurement inaccuracy at low impeller frequencies.

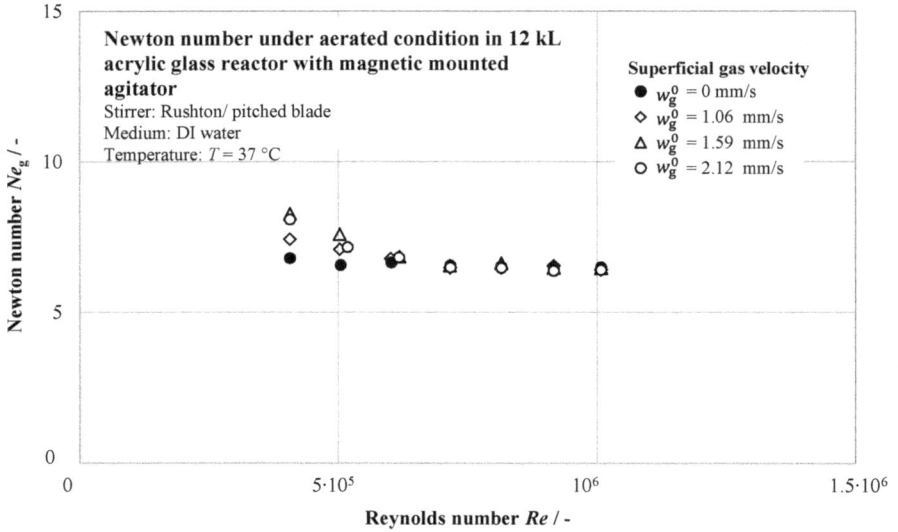

Figure 4-9: Influence of the superficial gas velocity on the Newton number in the large scale system

A measurement error is indicated by the fact that the idling torque cannot be determined. This needs to be determined iteratively at high Reynolds numbers, as described in Chapter 2.1. Especially at high gassing rates, there is a strong movement of the stirrer shaft, which can lead to a larger unknown torque. Alternatively, the flow around the stirrer blades can be changed by a high superficial gas velocity, so that no stationary flow is formed. This leads to a greater drag in the fluid and thus to an increase in the Newton number. With larger Reynolds numbers, the rotational speed of the stirrer predominates the flow and a stationary flow is formed again, resulting in a smaller Newton number.

Comparison of small scale and large scale system

In Figure 4-10, the results of the small scale and the large scale system are compared with the presented data from Zlokarnik. The unfilled symbols are literature data for the decrease of the Newton number with increasing Flow number for a Rushton turbine (diamond) and a pitched blade turbine (circles). Zlokarnik presented the influence of gas flow rate on the Newton number with the Flow number. He stated that at a certain Flow number the Newton number starts to decrease.

It can be seen that the combination of a Rushton turbine and a pitched blade turbine ($Ne = 6.5$), which were measured in the 12 kL reactor, is higher than the Newton numbers of the single Rushton turbine ($Ne = 4.9$) and the single pitched blade turbine ($Ne = 1.5$), which had been presented by Zlokarnik [Zlo03]. Furthermore, it can be seen that only a small range of Flow numbers had been investigated due to the limitations in mammalian cell cultivation.

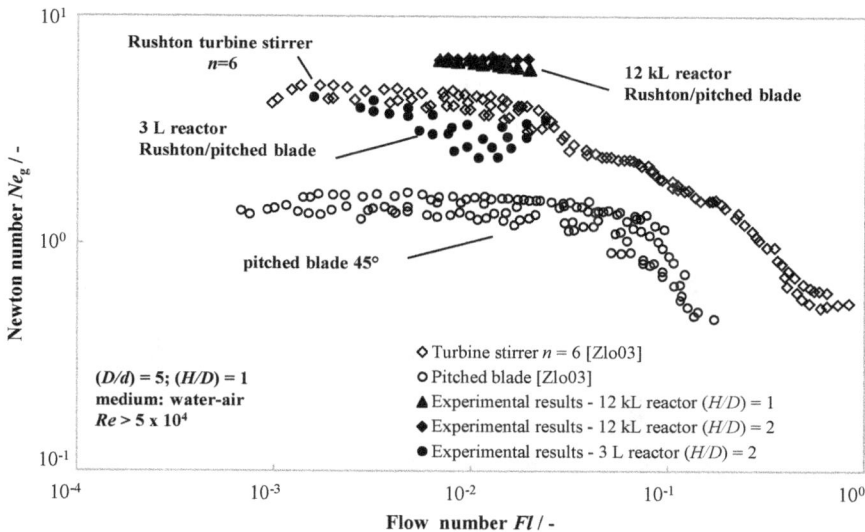

Figure 4-10: Newton number in aerated systems in dependency on the Flow number in the small scale system (filled circles) and the large scale system (filled diamonds and triangles) in comparison with literature data (empty symbols) [Zlo03]]

For the large scale reactor, the Flow number was limited to the value of $Fl = 0.007$ to $Fl = 0.022$. In this range the literature shows a reduction of the Newton number for the Rushton turbine (empty diamonds) with a level ratio of $(H/D) = 1$ from $Ne = 4.9$ to $Ne = 3.2$ which is equivalent to a reduction of 35 %. The investigations in the large scale reactor show a smaller influence on the Newton number which is furthermore dependent on the liquid height. For a level ratio $(H/D) = 1$ (filled triangles), a small influence can be seen. The Newton number decreases from $Ne = 6.6$ to $Ne = 5.8$, which is equivalent to a reduction of 12 %, whereas for a liquid level $(H/D) = 2$ (filled diamonds), no significant reduction of the Newton number is recognisable.

In the small scale system (filled circles) on the other hand, the influence of the Flow number on the Newton number is larger compared to the literature data. In the same range, the Newton number decreases from $Ne = 4.6$ to $Ne = 2.3$ (50 %) and is thus neither comparable to literature data nor to the large scale system with the same geometrical ratio.

The reason for the different influences of the gaseous phase on the power input lies in the fact that the surface tension and thus the sizes of the gas bubbles are not scalable. As a result, the bubbles in the large scale system are much smaller in relation to the stirrer blade size and thus to the formed vortices. The bubbles easily get trapped in the vortices and form the two cavities. In the small scale system, the cavities are much smaller than the bubbles, which are in the same size of the stirrer blade. Due to the surface tension it is not possible to form the same cavities as in the large scale system. Instead, the bubbles get trapped behind the stirrer blade and form the large cavities as seen

71

below. These cavities lead to better streamlines around the stirrer blade and thus to a reduced power input. It is also possible to form these cavities in the large scale system but much higher gas hold-ups are needed. It can be concluded that the reduction of the power input is decreasing with increasing reactor size for constant Flow numbers.

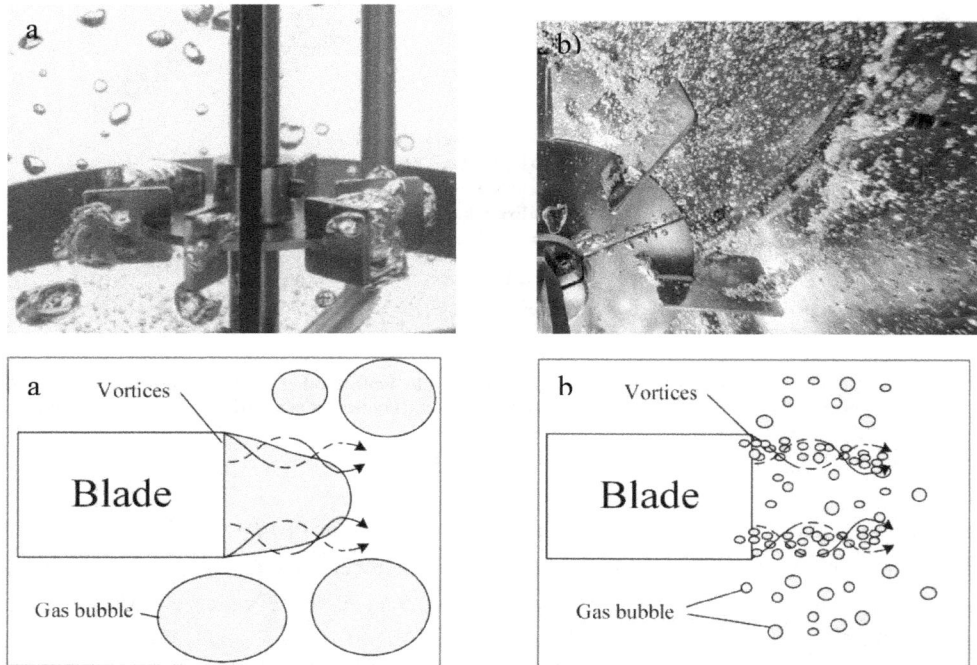

Figure 4-11: Ratio of gas bubbles to blade and vortex size in a small scale system with one large cavity (a) and a large scale system with two small gas filled vortices (b)

Summary

The investigation shows that the scale-up of the power input is only possible for a single phase system. With the help of the flow number, the influence on the Newton number can be roughly estimated, but not exactly predicted, since other parameters such as surface tension must be included on different size scales. It could be shown that the influence in the large scale system can be neglected and that the stirred power input can be estimated via the unaerated Newton number for all cases. In the small scale system on the other hand, the influence needs to be taken into account.

4.2 Bubble size distribution, gas hold-up and specific surface area in small and large scale system

In fermentation processes, usually, the needed oxygen is provided by classical aeration of the liquid. For a profound understanding of the mass transfer processes, the bubbly flow needs to be characterised in terms of the bubble size distribution as well as the gas hold-up. Beside of the mass transfer, the bubble size distribution has a strong effect on hydrodynamic and thus on the flow pattern and on the mixing time.

In this chapter the bubble size distribution and the influence on the flow pattern are characterised for the lab scale and the industrial scale reactor and the main differences are identified.

4.2.1 Investigation of bubble size distribution

Small scale system

In a bubbly flow, a wide range of bubble sizes are present and a mean diameter needs to be defined for a representative description. For mass transfer performances typically the Sauter mean diameter is used, because it combines the volume and the surface area of the whole gaseous phase. In Figure 4-12, the Sauter mean diameter d_{32} is presented as a function of the superficial gas velocity for the small scale system.

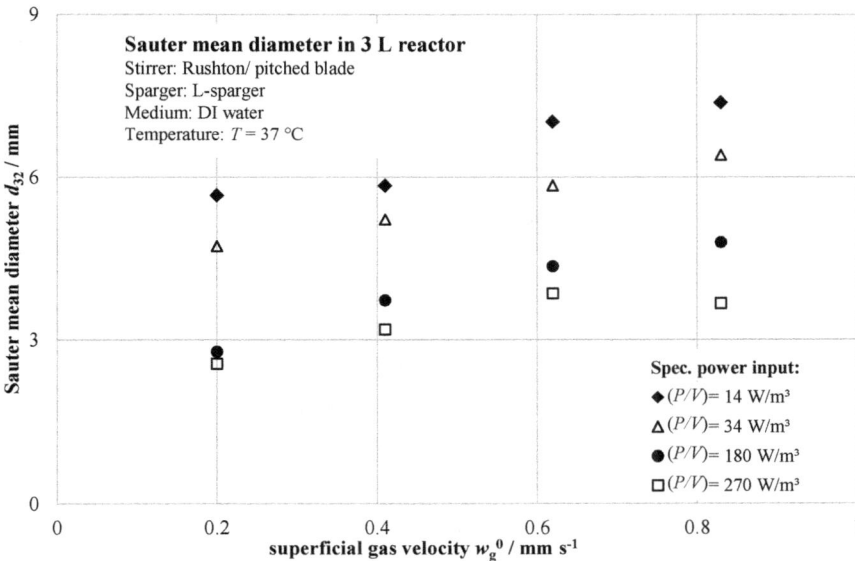

Figure 4-12: Sauter mean diameter in 3 L glass reactor as a function of the superficial gas velocity

It can be seen that the Sauter mean diameter is increasing with increasing superficial gas velocity which is in consistence to the literature survey. Due to the higher gas flow rate, the bubbles are growing faster and thus are larger when they detach from the nozzles. Furthermore, due to the increased amount of gas in the system, fewer bubbles are getting dispersed by the stirrer and the probability of coalescence is increasing which leads to a larger Sauter diameter. The influence on the stirring can also be seen in Figure 4-12. With increasing power input the bubble sizes are decreasing for all investigated gas flow rates. Two basic mechanism of the liquid motion are influencing the bubble size distribution. Due to the movement of the fluid, induced by the stirrer, the bubbles detach more quickly from the orifices than in liquids without movement. As a result, the Sauter mean diameter decreases with increasing stirrer frequency and thus with increasing power input. The second mechanism is the dispersion of the bubbles in the stirrer area. This break-up is more effective for higher stirrer frequencies and leads to smaller bubbles for higher specific power input.

Beside the Sauter diameter, the inhomogeneity of the bubble size distribution is an important factor for the process. One way for the description of the inhomogeneity is the width of the bubble size distribution. The width $\Delta d_{10,90}$ is the width where only the lower and upper 10 % of the bubble size distribution are excluded. The smaller $\Delta d_{10,90}$, the more bubbles have the same size and thus, the more homogeneous the bubble size distribution is. This can be shown schematically in Figure 4-13 for the 3 L reactor and a specific power input of $(P/V) = 170$ W/m³. As the gassing rate increases, the width of the distribution also increases. This means that with increasing gassing rate the bubble flow becomes more inhomogeneous.

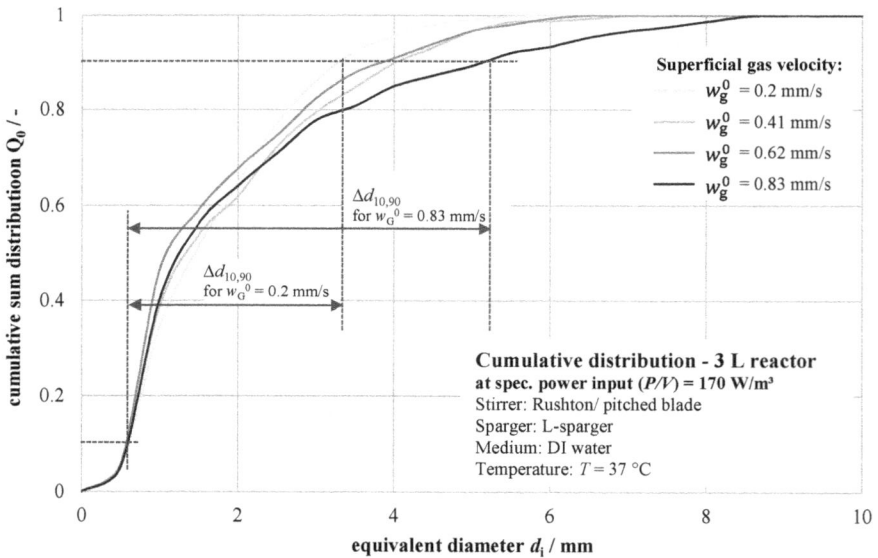

Figure 4-13: Cumulative bubble size distribution in the 3 L reactor with a power input $(P/V) = 170$ W/m³

In Figure 4-14 the inhomogeneity is presented as a function of the superficial gas velocity for different specific power inputs. It can be seen that an increase of the power input leads to a smaller distribution and thus to a more homogeneous bubbly flow. For high power input, the gas flow rate has almost no influence on the bubble size distribution, because the whole gaseous phase gets dispersed in the stirrer region. Furthermore, the high liquid velocity inhibits the coalescence of the bubbles.

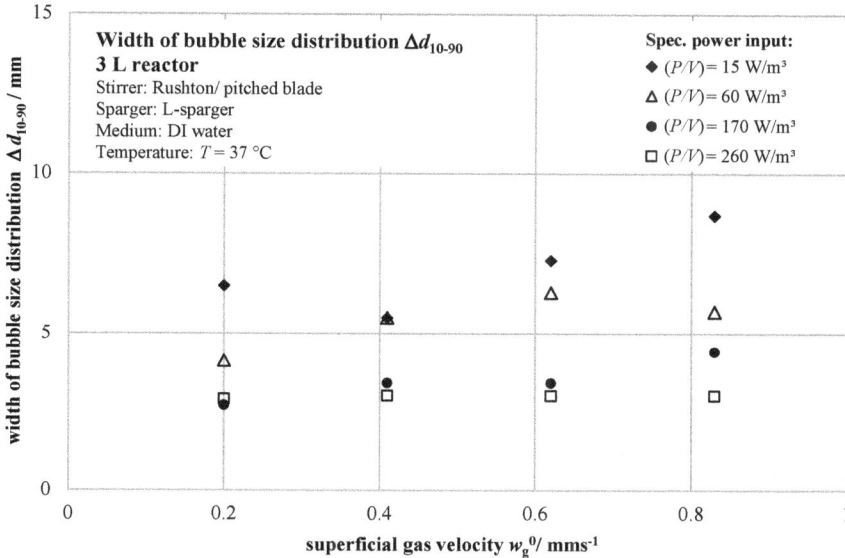

Figure 4-14: Width of bubble size distribution in 3 L reactor

This is different for a lower power input. In contrast to a high power input, only parts of the gaseous phase get dispersed whereas some of the primary bubbles are passing the region and remain at the same size. Additionally, due the smaller liquid velocities, more coalescence can occur, which leads to larger bubbles in the system and thus to a broad bubble size distribution. This effect gets stronger for higher gas flow rates, which is the reason for the increased inhomogeneity with increasing gas flow rate.

Large scale system

In Figure 4-15, the Sauter mean diameter in the 12 kL acrylic glass reactor is presented as a function of the superficial gas velocity for two different specific power inputs. Similar to the small scale system, the superficial gas velocity has only a small influence on the Sauter mean diameter. For a low energy input, the diameter is increasing from $d_{32} = 4$ mm to $d_{32} = 6$ mm. Similar to the small scale system, the power input is not sufficient to disperse the whole gaseous phase, which leads to larger bubbles and thus to a larger Sauter mean diameter. For high energy input, the gas flow rate has almost no influence on the bubbles. Here the whole gaseous phase gets dispersed in the impeller region.

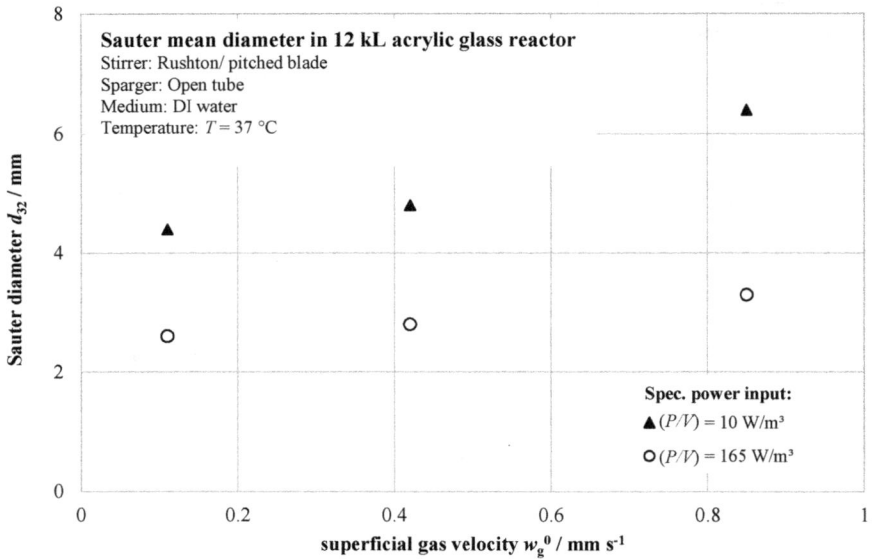

Figure 4-15: Sauter diameter in the acrylic glass reactor as a function of the power input for two different superficial gas velocities

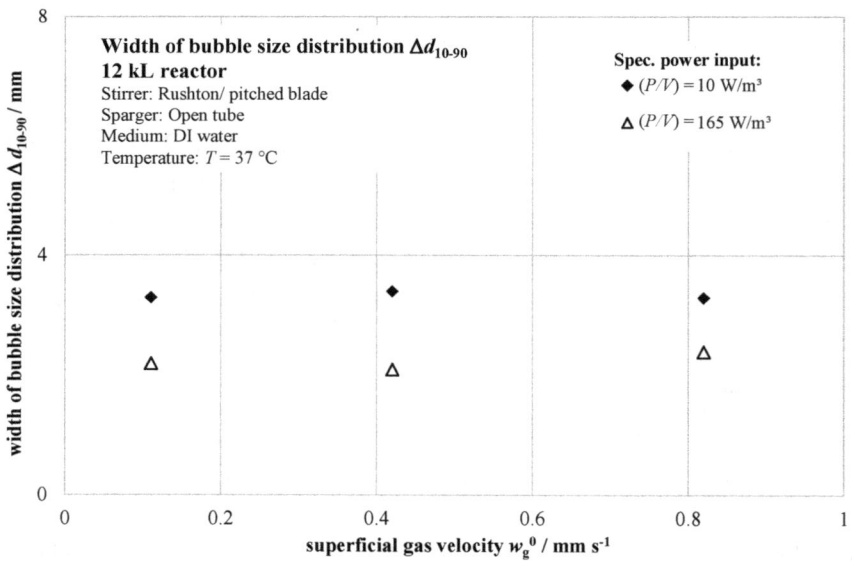

Figure 4-16: Width of bubble size distribution in 12 kL acrylic glass reactor

In Figure 4-16, the inhomogeneity indicated by the width of the bubble size distribution is presented as a function of the superficial gas velocity. As for the Sauter mean diameter, the width is decreasing with increasing stirrer frequency. The gas flow rate on the other hand seems to have no significant influence on the width of the bubble size distribution for the large scale system.

The results of the distribution are in conflict to the results of the Sauter mean diameter in Figure 4-15, where, for the lower specific power input of $(P/V) = 10$ W/m³, an increase in the gas flow rate leads to an increase in the Sauter mean diameter. This can have two reasons. First, with increasing gas flow rate, the diameter of all bubbles in the system increases almost equally. But looking at the cumulative distribution in Figure 4-17 shows that the distributions of the different gas flow rates are almost identical. The second explanation might be that the major amount of the bubbles still have the same sizes as for a lower gas flow rate because the main dispersion takes place by the impeller and therefore is unaffected by the amount of gas. But with increasing gas flow rate some parts of the gas, which cannot be dispersed by the impeller, are rising as large umbrella-shaped bubbles. Compared to the large total number of dispersed gas bubbles, the small number of large umbrella bubbles is not taken into account with this method and thus the bubble size number distribution seems to be almost the same.

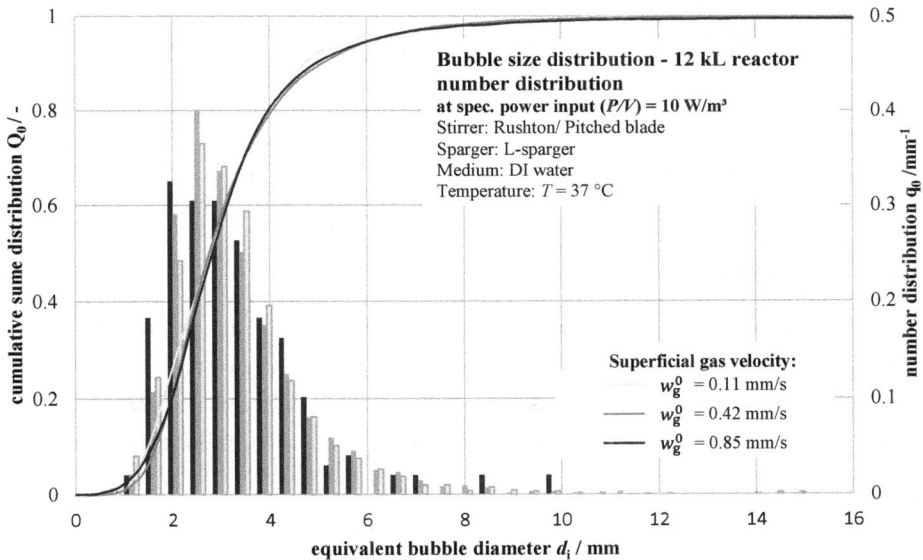

Figure 4-17: Bubble size distribution in the 12 kL reactor for the power input $(P/V) = 10$ W/m³

For the calculation of the Sauter mean diameter, both the surface and the volume of each bubble are included. This means that already one large bubble has a strong influence on the determined value of the Sauter mean diameter. Thus, it is useful also to take the volume distribution q_3 of the bubble size distribution into account (see Figure 4-18). Since the diameter of a bubble is influencing the volume to the power of three, only one bubble can lead to large difference in the volume distribution q_3

while the standard number distribution q_0 is not affected. As a characteristic bubble diameter of the volume distribution, the volumetric mean diameter d_{m3} can be used, which is the diameter where the volume sum function is equal to $Q_3 = 0.5$. It can be seen in Figure 4-18 that the mean diameters d_{m3} is changing for different superficial gas velocities, while the mean diameter d_{m0} in Figure 4-17 is almost constant for all superficial gas velocities.

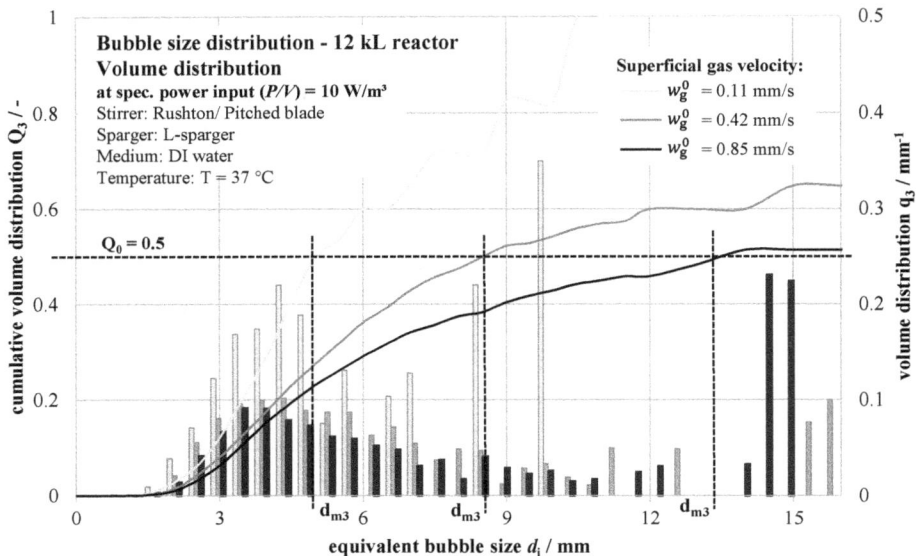

Figure 4-18:   Volume distribution in the 12 kL reactor for $(P/V) = 10$ W/m³

Comparison of small scale and large scale system

The aim of the investigation of the bubble size distribution is the definition of better and more reliable scale-up rules. While the power input can easily be scaled, this is almost impossible for a bubble size distribution. Nevertheless, it is important to understand the basic differences between the different scales. The Sauter mean diameters for both systems show similar dependency on the gas flow rate.

The influence of the power input on the Sauter mean diameter is stronger in the large scale system. However, the bubble sizes are at the same magnitude in both scales. To describe the heterogeneity of a bubble size distribution, the width can be used. The widths at the scales are also of the same order of magnitude. However, it is much broader in the small scale system, compared to the large scale. Furthermore, the superficial gas velocity seems to have no influence on the heterogeneity in the large scale system. This leads to the first assumption that a stronger heterogeneity can be apparent in the small scale system, which is against the expectations.

However, a few large bubbles have no influence on the number distribution, but can have a strong influence on the hydrodynamic. Thus for the hydrodynamic the volume distribution also needs to be taken into account.

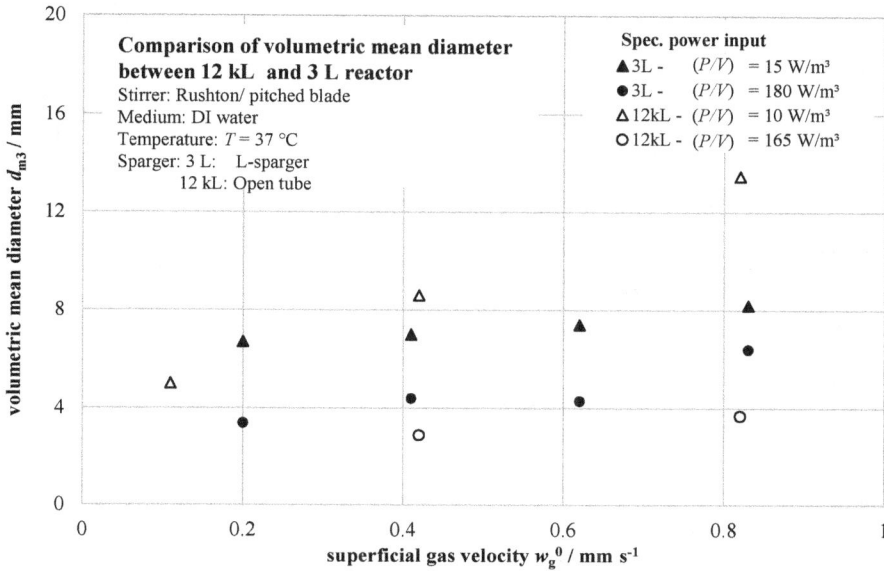

Figure 4-19: Comparison of the volumetric mean diameter for the small scale and the large scale reactor

In Figure 4-19, the volumetric mean diameter d_{m3} for the small scale and the large scale system are compared as a function of the superficial gas velocity for different power inputs. In this comparison, the different mechanisms at the two scales become particularly clear. While the mean diameter in the large scale system increases significantly at low power input with increasing gas flow rate, it remains almost constant at high energy input. In the small scale system, on the other hand, aeration has no significant influence on the mean size at either high or low energy input.

This strong difference can be explained with the different dispersion mechanisms at the two scales. In the small scale system, the first dispersion is already happening during the formation of the bubbles at the sparger. The sparger is equipped with nozzles of a diameter $d_n = 1$ mm that leads to small bubble diameters. In the stirrer region, these bubbles are either getting dispersed further or are just passing the region without a break-up. This leads to a wide bubble size distribution q_0.

In the large scale system on the other hand, the open tube sparger does not lead to a proper dispersion of the gaseous phase. Large umbrella-shaped bubbles are leaving the sparger and reaching the impeller region. Here, the gaseous phase is trapped by the vortex system of the impeller blades (see Figure 4-20) and are getting dispersed as small bubbles over the whole cross section of the reactor. The amount of gas that can be dispersed is dependent on the stirrer frequency. For high stirrer frequencies, the whole gaseous phase gets trapped and dispersed by the vortex system. For lower stirrer frequencies, only a certain amount can be dispersed. When the gas flow rate exceeds this volume, some parts of the gas are passing the impeller region without being dispersed.

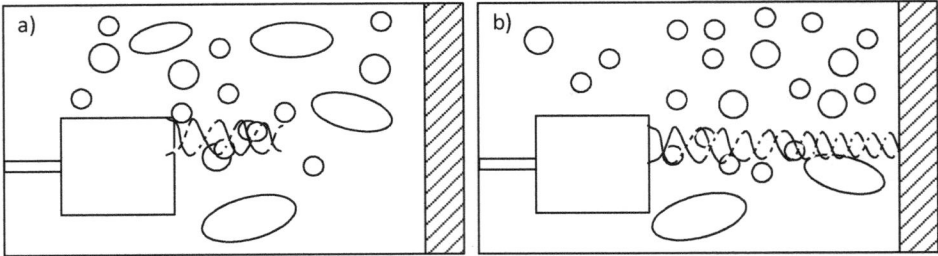

Figure 4-20: Dispersion mechanism by vortices behind the stirrer blade – a) low power input allows large bubbles to pass the stirrer (heterogeneous bubbly flow) – b) strong momentum of impeller leads to a sufficient break up of the bubbles (homogeneous bubbly flow)

4.2.2 Investigation of the gas hold-up and specific surface area

Small scale system

With the amount of gas bubbles in the reactor the gas hold-up can be determined in the small scale reactor. The results are presented in Figure 4-21. As expected, the gas hold-up increases with increasing superficial gas velocity, as more gas is present in the system due to the higher gassing rate. Especially at a high gassing rate it becomes clear that the gas hold-up increases with increasing energy input. Due to the higher stirrer frequency, the gas gets dispersed more efficiently which leads to an overall decrease of the bubble sizes. Smaller bubbles rise with a lower velocity which results in a higher gas hold-up. For low superficial gas velocities, this effect is less visible because measurements at small gas contents are extremely difficult. Because the bubble size distribution decreases with increasing power input for both large and small superficial gas velocities, it can be assumed that the gas hold-up also increases here.

With the Sauter mean diameter and the gas hold-up, the specific surface area at which the mass transfer takes place can be calculated with the formula $a = \frac{6 \cdot \varepsilon_g}{d_{32}}$. The results for the specific surface area are presented in Figure 4-22 as a function of the specific power input and the superficial gas velocity. The specific surface area is directly proportional to the gas hold-up resulting in the same trend depending on the superficial gas velocity and the power input. Furthermore, the specific surface area is reciprocally dependent on the Sauter mean diameter. Since it decreases with increasing power input, as presented in chapter 4.2.1, the trend of the specific surface area is intensified.

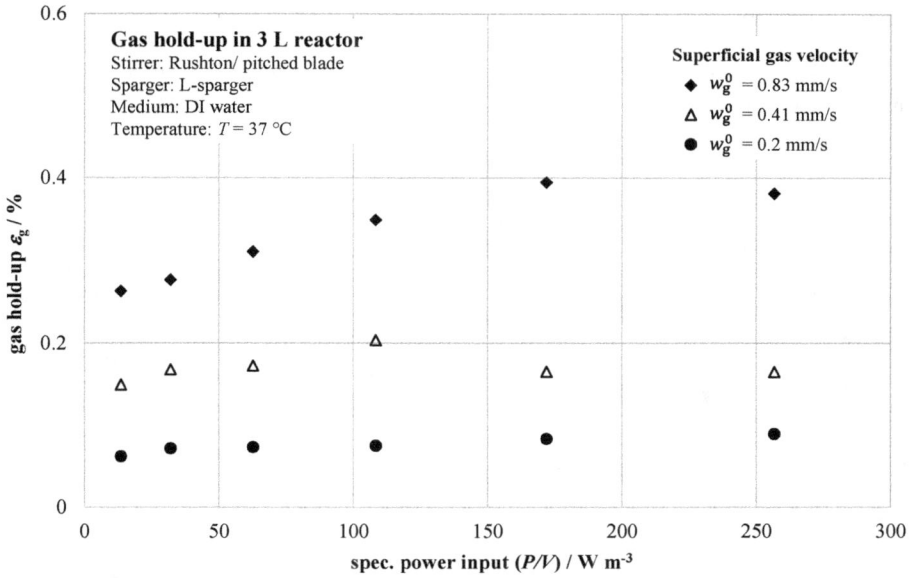

Figure 4-21: Gas hold-up in 3 L glass reactor as a function of the specific power input

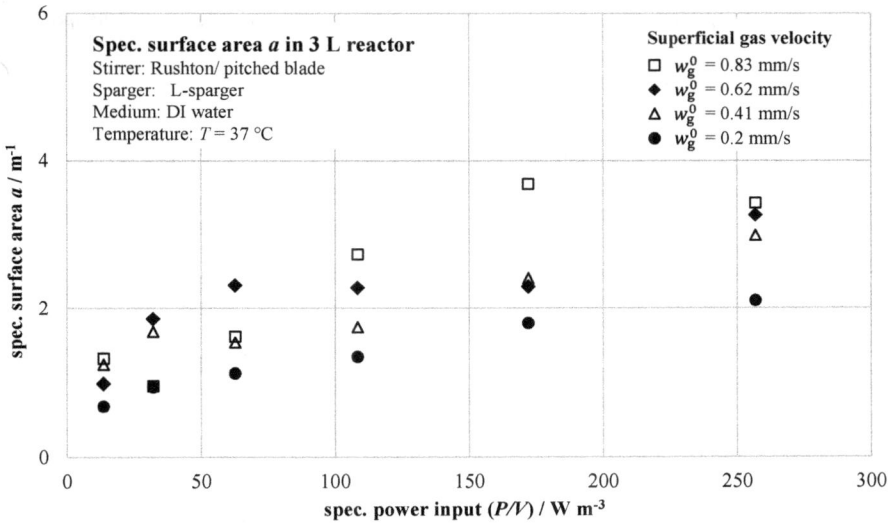

Figure 4-22: Specific surface area as a function of the power input for three different superficial gas velocities

Large scale system

In the acrylic glass reactor, the gas hold-up can be calculated with the variation of the liquid height during aeration (see chapter 3.3.3). The results are presented in Figure 4-23 as a function of the specific power input.

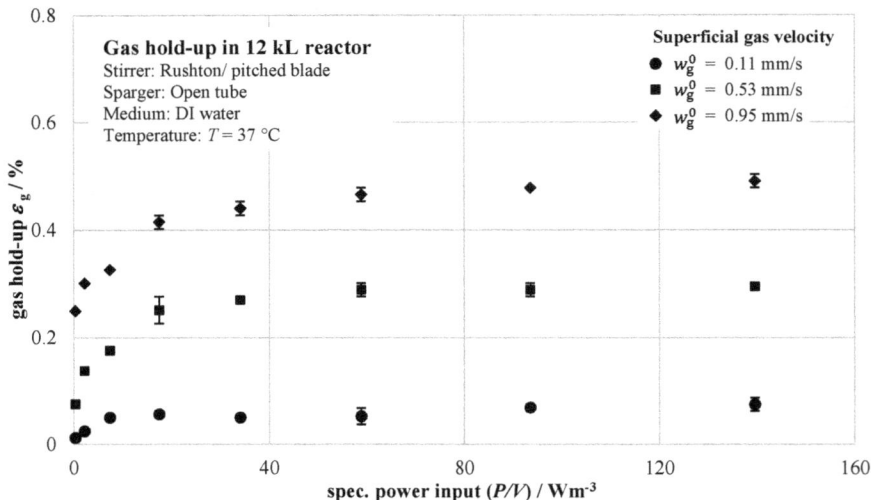

Figure 4-23: Gas hold-up within the acrylic glass reactor with $V_{fill} = 12\,000$ L as a function of the specific power input

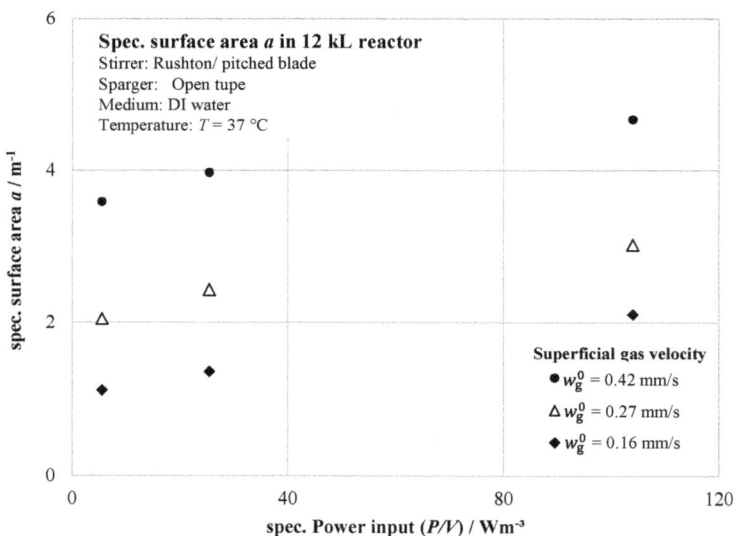

Figure 4-24: Specific surface area in the acrylic glass reactor with $V_{fill} = 12\,000$ L as a function of the power input for three different superficial gas velocities

It can be seen that in the range of small power input the gas hold-up increases with increasing power input. When a certain power input is reached, a further increase of power input does not lead to a further increase of the gas hold-up. This can be explained with the dispersion effect of the Rushton turbine. For small energy input, only a small amount of gas can be captured in the vortices behind the blades. The gas that exceeds this amount passes the stirrer region as large bubbles. When the stirrer frequency is increased, a larger amount of gas can be dispersed which leads to smaller bubbles with decreased rise velocities. At a certain point, the stirrer is not able to disperse the bubbles further. Only a further recirculation of the gas phase can lead to a further increase of the gas hold-up.

With the information of the Sauter mean diameter and the gas hold-up, the specific surface area can be calculated. The results can be seen in Figure 4-24. As in the small scale system, the specific surface area is increasing with both an increasing power input and an increasing superficial gas velocity.

4.2.3   Influence of the bubbly flow on the flow pattern

The content of this subchapter correspond in large parts to the published article [Ros18].

Inhomogeneity of the bubble size distribution is an important parameter to consider for the design of large scale reactors. In the large scale reactor, a much wider bubble size distribution can occur, compared to the small scale system. Furthermore, the bubble sizes are very small in relation to the diameter of the reactor. This can lead to an inhomogeneous bubble distribution over the cross section, with areas containing very large bubbles, areas with very small bubbles and areas where only few bubbles are present. In laboratory reactors, the bubbles are significantly larger in relation to the diameter of the reactor, preventing these heterogeneities from occurring. This differences can schematically be seen in Figure 4-25.

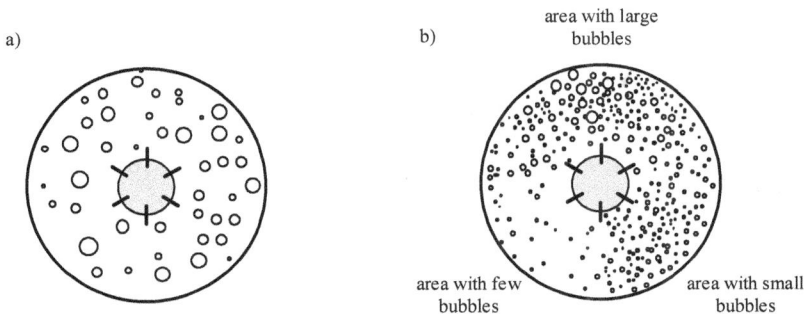

Figure 4-25:   Distribution of gaseous phase in a 3 L reactor a) and a 12 kL large scale reactor b)

For this reason, only the inhomogeneity of the large scale system has been investigated further. The different flow regimes can be determined quantitatively with the transition of loading and flooding and thus with the critical Flow number Fl_c. To determine the transition between loading and flooding in the acrylic glass reactor, the gas hold-up has been measured for a wide range of parameters.

In Figure 4-26, the gas hold-up ε_G is shown as a function of the Froude number for three different exemplarily gas flow rates. As expected, with increasing superficial gas velocity and increasing stirrer frequency, the gas hold-up increases. However, the transition from loading to flooding can be determined by a change in the slope of the graphs (red circles). For higher stirrer frequencies the gas hold-up remains almost constant because of the efficient dispersion of the gaseous phase by the Rushton turbine. In this region a homogeneous bubbly flow can be observed. With decreasing stirrer frequency a sudden drop of the gas hold-up can be found because the momentum induced by the turbine is not sufficient anymore to disperse the gaseous phase completely. Bigger bubble and bubble agglomerates appear, which are passing the turbine with high rise velocity, leading to a drop in gas hold-up and the transition to flooding. This can be visualised clearly within the acrylic glass reactor at the upper end of the rotating stirrer axis.

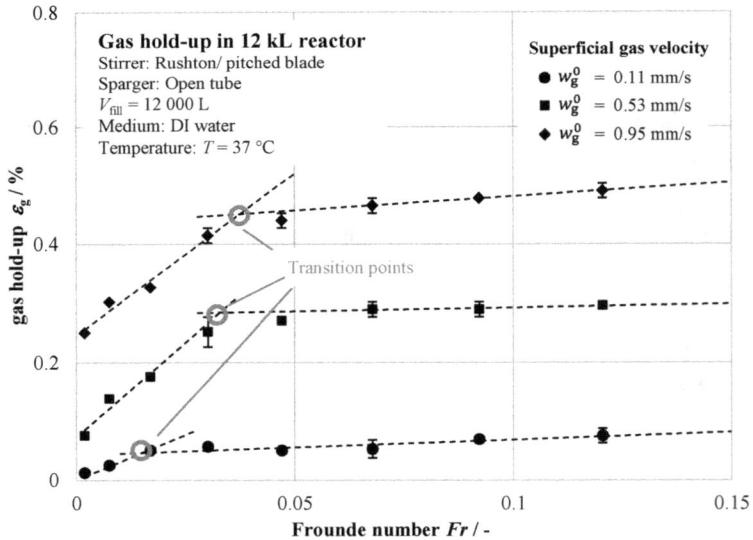

Figure 4-26: Gas hold-up in acrylic glass reactor with V_{fill} = 12 000 L as a function of the specific power input with the marked transition points [Ros18]

Figure 4-27 a) shows the heterogeneous regime with a very broad bubble size distribution and gas hold-up distribution, whereas Figure 4-27 b) shows the same area at a higher Froude number and lower Flow number that causes a homogeneous flow regime with a narrow bubble size distribution as well as a homogeneous distribution of the gaseous phase over the reactor cross section.

In Figure 4-28 the transition points detected in Figure 4-26 are compared with literature data. As can be seen, a good agreement between the experimental data and the correlation of Mikulcova (Formula (2.20)) is archived.

From this investigation it can be concluded that the transition between loading and flooding for the industrial scale acrylic glass reactor is in the same range as presented by Mikulcova and can be characterised with the critical Flow number

$$Fl_c = 0.42 \cdot Fr \qquad\qquad (4.1).$$

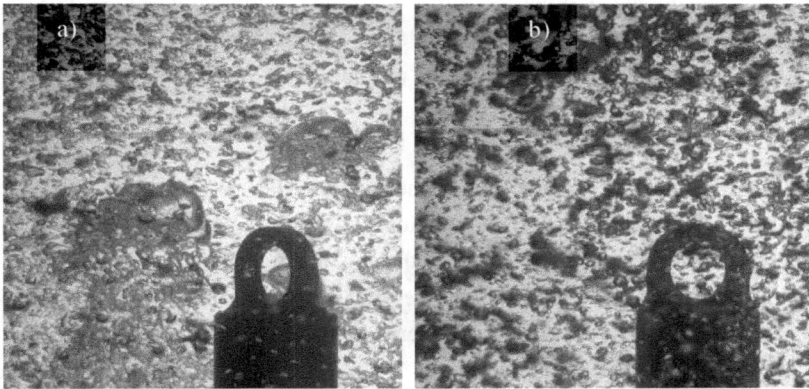

Figure 4-27: **Exemplary heterogeneous bubbly flow a) and homogeneous bubbly flow b)**

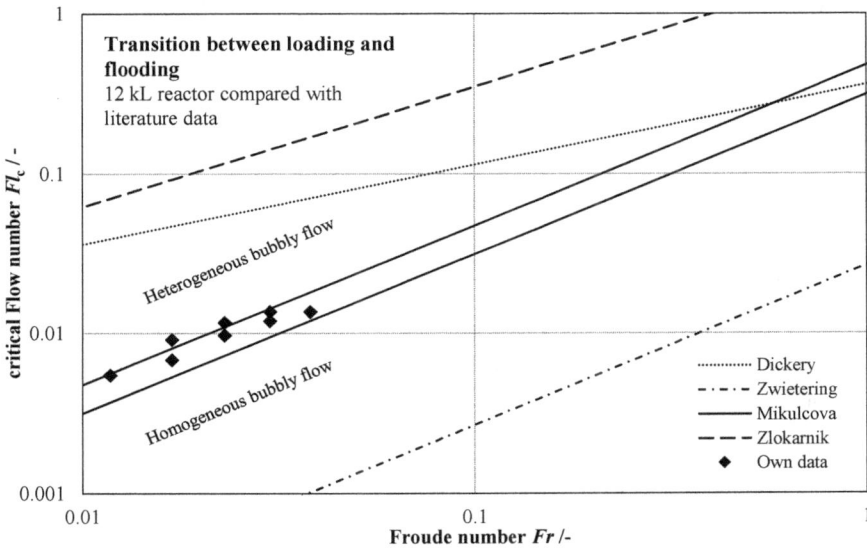

Figure 4-28: **Transition between homogeneous and heterogeneous flow regime [Ros18]**

85

4.3 Influence of flow pattern on mixing time

The content of this chapter correspond in large parts to the published article [Ros18].

The following section provides the experimental results for mixing times that have been achieved in the 12 000 L acrylic glass reactor by means of the decolouration method. First, the results without aeration will be presented. Even though, the unaerated processes play a minor role in the fermentation, only for these sufficient literature data is available to perform a good validation. Furthermore, aerated mixing is often described as a function of the unaerated mixing. Therefore, deep knowledge of the unaerated mixing time is also of great importance. The results of the aerated mixing time are presented following those of the unaerated mixing time.

4.3.1 Mixing time for single phase stirring

In Figure 4-29, the mixing time for the acrylic glass reactor is presented as a function of the Reynolds number for different impeller combinations and level ratios (*H/D*). As expected, the mixing time decreases with increasing stirrer frequency due to the higher energy dissipation rate and induced turbulence for all cases. During the measurements, a very high spreading of the measurement results was noticed, which is reflected in the error bars. At low stirrer frequencies, the deviation between the single measurements are 75 %. At higher stirrer frequencies, the deviations have been reduced to 70 %, but still are relatively high.

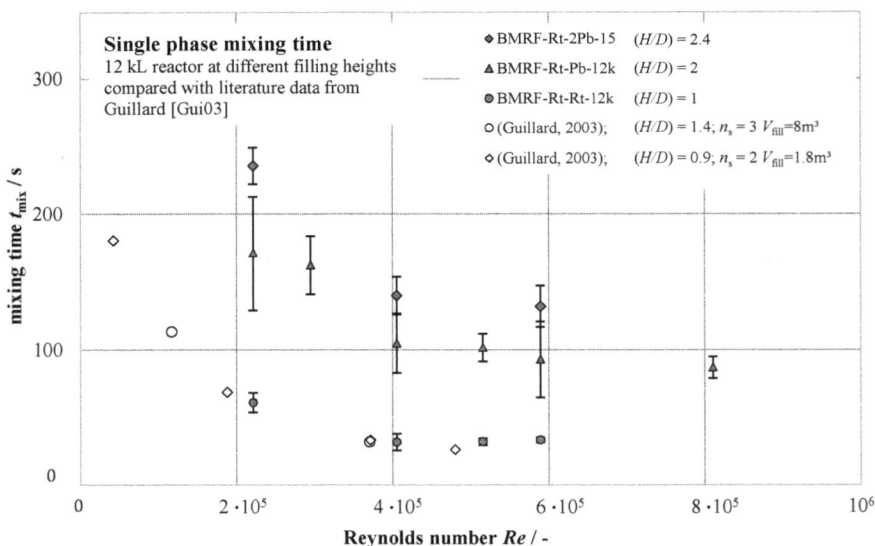

Figure 4-29: Mixing time in dependency on the Reynolds number, filling level and stirrer combination

86

	V_{fill} = 12 000 L (P/V) = 170 W/m³ w_G^0 = 0 mm/s					V_{fill} = 12 000 L (P/V) = 10 W/m³ w_G^0 = 0 mm/s				
	5	4	3	2	1	5	4	3	2	1
0 s										
10 s										
20 s										
30 s										
40 s										
50 s										
60 s										
70 s										
80 s										
90 s										
100 s										
110 s										
120 s										
130 s										
140 s										
150 s										
160 s										
170 s										
180 s										
200 s										
210 s										
220 s										

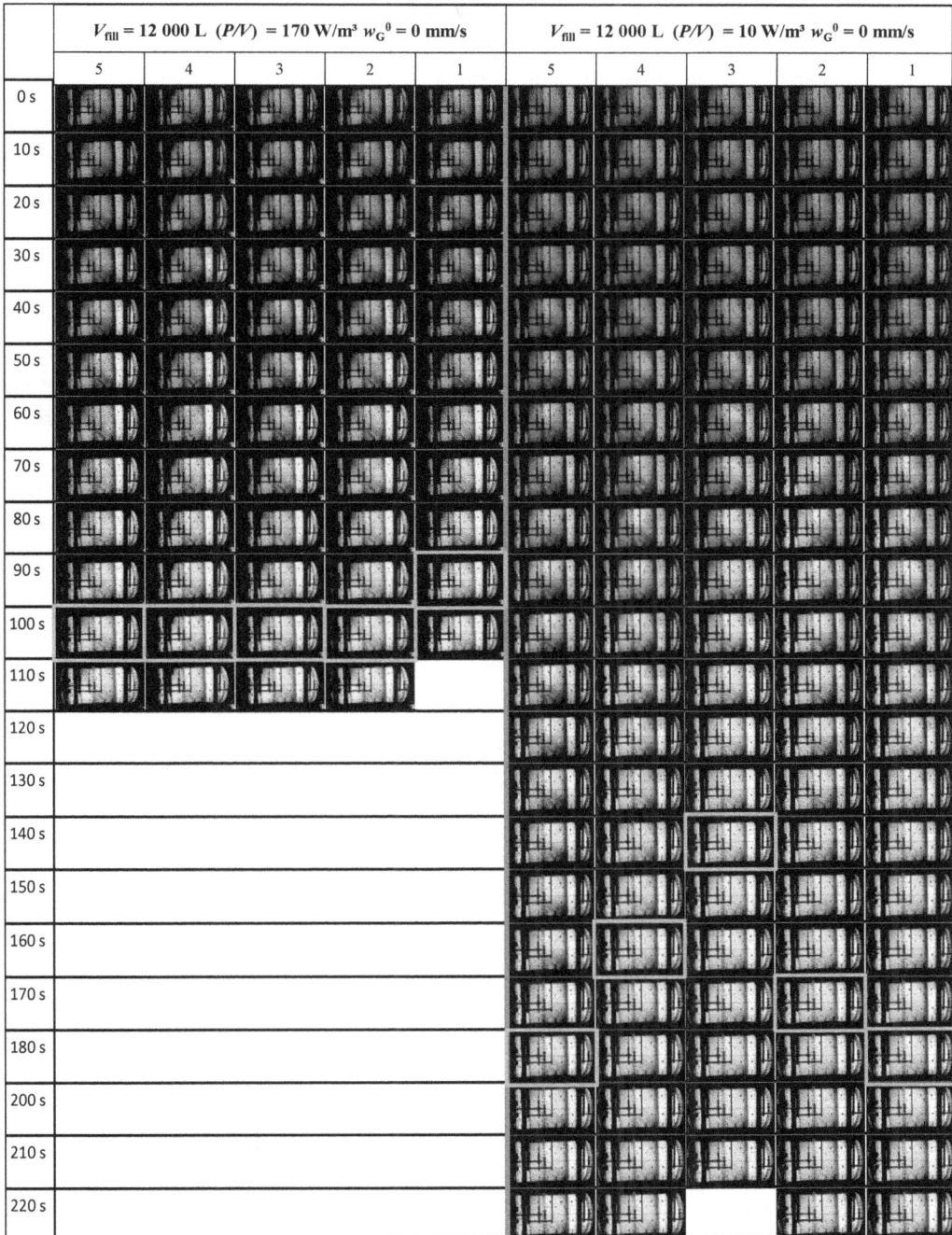

Figure 4-30: Single measurements of the mixing times without aeration for two different specific power inputs

A closer look reveals that, despite the fact that the acid addition is identical in every experiment, the decolouration process differs significantly between each measurement. Five exemplary single decolouration processes can be seen in Figure 4-30 for two different specific power inputs.

In some experiments the decolouration occurred first in the upper part of the reactor (e.g $(P/V) = 10$ W/m³, exp. 3 in Figure 4-30). In other experiments (e.g $(P/V) = 10$ W/m³, exp. 1 in Figure 4-30) the medium decolourised first in the lower area. These different decolouration behaviours indicates that only a semi-stationary process with periodically occurring flows has developed. These strong deviations were no longer apparent in the aerated state. Thus, this thesis does not go into more detail here.

A comparison of the results with literature data is difficult, because publications that are dealing with experimental results on this large scale are rare. In most cases different geometries or stirrer combinations have been used. Investigations on similar scale have been published by Guillard and Trädgårdh [Gui03]. They investigated the mixing time on two different scales ($D_1 = 1.88$ m/ $V_{fill} = 8$ m³ and $D_2 = 1.4$ m/ $V_{fill} = 1.8$ m³) with multiple impeller systems (3 Rushton turbines and 2 Rushton turbines respectively) by measuring the local conductivity. The comparison shows that the results obtained in the acrylic glass reactor with a filling volume of $V_{fill} = 6$ m³ ($H/D = 1$, filled circle) are in a good agreement with the measurements by Guillard for a filling volume of $V_{fill} = 8$ m³ (empty circle) and $V_{fill} = 1.8$ m³ (empty diamond). With increasing volume and (H/D) respectively, the mixing time increases.

To compare these results with common correlations, the dimensionless mixing time is used. In Figure 4-31 the dimensionless mixing time $t_{mix} \cdot n$ is presented for three different level ratios. For a better comparison between the different filling volumes, the dimensionless mixing time is plotted as a function of the power input instead of the Reynolds number.

It can be seen that the mixing time increases with increasing level ratio for the same power input. This also applies to the volume of $V_{fill} = 15$ m³ with (H/D) = 2.5, where an additional stirrer has been installed to support axial mixing in the upper part of the reactor. By means of the investigations, the following correlation

$$n \cdot t_{mix} \sim \left(\frac{H}{D}\right)^{-1.6}$$

for the influence of the level ratio could be found.

In addition to the influence of the filling level, Figure 4-32 also clearly shows that the dimensionless mixing time, despite many publications, also depends on the power input and thus on the Reynolds number. Most authors predict a constant dimensionless mixing time within the turbulent regime of Reynolds numbers $Re > 10^4$ which is also reflected in the correlations in Table 2-3, where $n \cdot t_{mix}$ is only a function of the impeller type and geometry of the reactor.

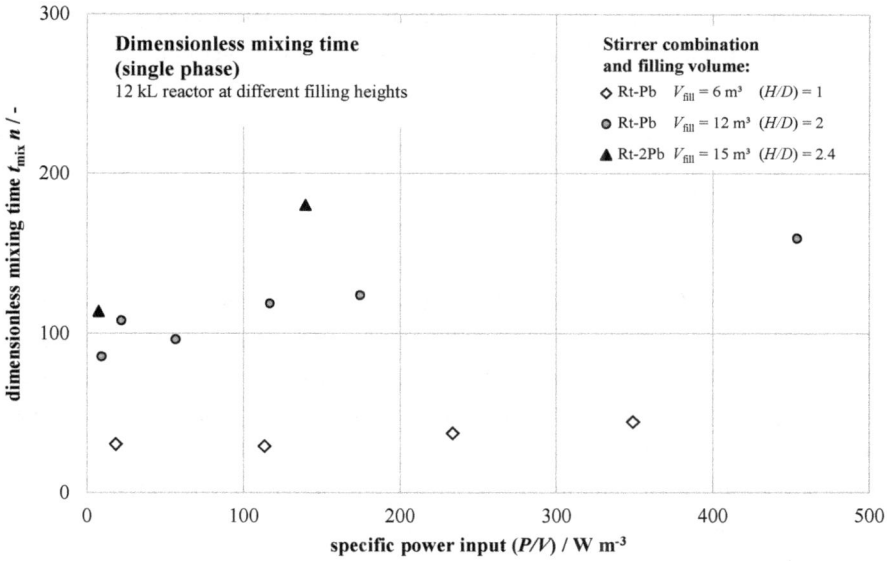

Figure 4-31: Dimensionless mixing time as a function of the specific power input for different level ratios

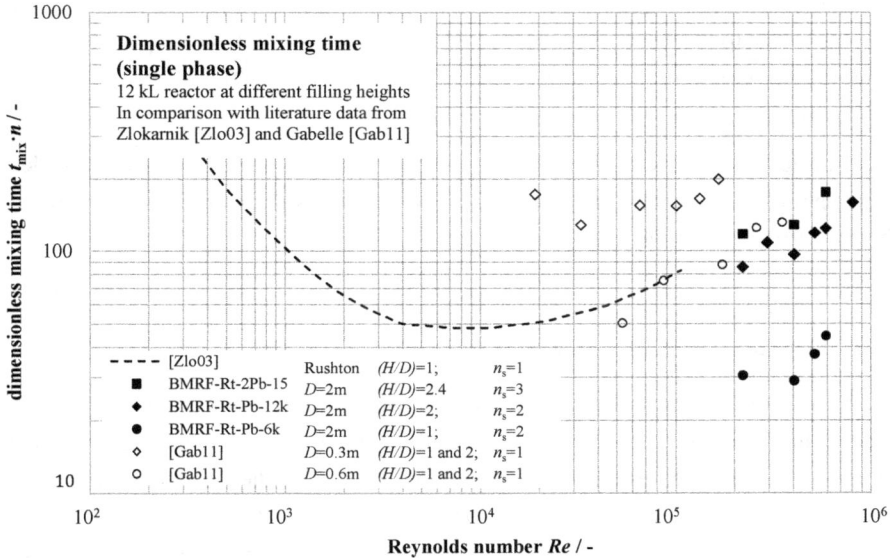

Figure 4-32: Dimensionless mixing time in dependency on Reynolds number for different level ratios (filled symbols) in comparison with literature data (empty symbols)

The dimensionless mixing times for the acrylic glass reactor with the three stirrer combinations (Rt-2Pb-15kL/ Rt-Pb-12kL/ Rt-Pb-6kL) from Figure 4-31 are plotted in Figure 4-32 as a function of the

Reynolds number. It can be seen that for $Re > 10^4$, the dimensionless mixing time increases with increasing Reynolds number. For all three combinations, a similar trend of the dimensionless mixing time in dependency of the Reynolds number can be noticed:

Rt-Pb-6kL: $t_{mix} \cdot n \sim Re^{0.33}$; *(H/D) = 1*

Rt-Pb-12kL: $t_{mix} \cdot n \sim Re^{0.47}$; *(H/D) = 2*

Rt-2Pb-15kL: $t_{mix} \cdot n \sim Re^{0.41}$; *(H/D) = 2.4*

Similar experimental results with comparable trends have been reported by Gabelle et al.in 2011 [Gab11] even for reactors of smaller volume but for high Reynolds numbers. Zlokarnik already mentioned in 1967 that the mixing time for very high Reynolds numbers tends to become constant because it is only limited by the constant diffusion in this regime [Zlo67b]. This indicates that the dimensionless mixing time needs to increase at a certain point. The dotted line in Figure 4-32 shows this trend of the dimensionless mixing time of a Rushton turbine proposed by Zlokarnik [Zlo67b]. This trend can be found in the data from Gabelle as well as in the data from the industrial scale acrylic glass reactor.

Despite the small amount of measuring points, the trend of single phase mixing makes clear that the approximation of a constant dimensionless mixing time needs to be treated with caution for large scale reactors. Due to the large dimension of the stirrer, the Reynolds number is much higher even for very low stirrer frequencies. Whereas in small scale reactors at moderate Reynolds numbers the assumption of a constant dimensionless mixing time is useful, in large industrial scale reactors an error of at least 20 % needs to be taken into account.

However, the acrylic glass reactor enables a detailed analysis of mixing times in dependency of the impeller combination, geometry and reactor aspect ratios and will provide data for more reliable and transferable correlations in the future.

4.3.2   Mixing time for aerated stirring

In a second step the influence of the gaseous phase on hydrodynamics and mixing times has been investigated. In Figure 4-33, the mixing time for different gas flow rates can be seen for the standard configuration. The lines represent the correlation by Alves with the fitted constants *A*, *B* and *C*. It can be seen for the experiments and for the correlation that the mixing time decreases with increasing stirrer speed for the unaerated condition (diamond). Furthermore, it becomes clear that gassing generally has a positive effect on the mixing, which is, however, differently intense within the investigated area. In the range of high Reynolds numbers, the bubbles are rising homogeneously and lead locally enhanced mixing which only has a small influence on the overall mixing time. Furthermore, it can be seen that for very small ($Re < 10^5$) and very large ($Re > 4 \cdot 10^5$) Reynolds numbers, the dependency on the stirrer speed is analogous to the unaerated condition.

For mammalian cell cultivation processes however, Reynolds numbers between $Re = 10^5$ and $Re = 3 \cdot 10^5$ are relevant. In this range the mixing time behaves contrary to the other ranges, where an increase of the stirrer speed leads to a reduction of the mixing time. This effect has already been described in literature, but so far there are almost no published correlations for this case. Most authors describe the influence of aeration by a constant factor that does not reflect the trend found in these experiments. Only the correlation of Alves (eq. (2.44)) provides good agreement in the relevant Reynolds numbers range.

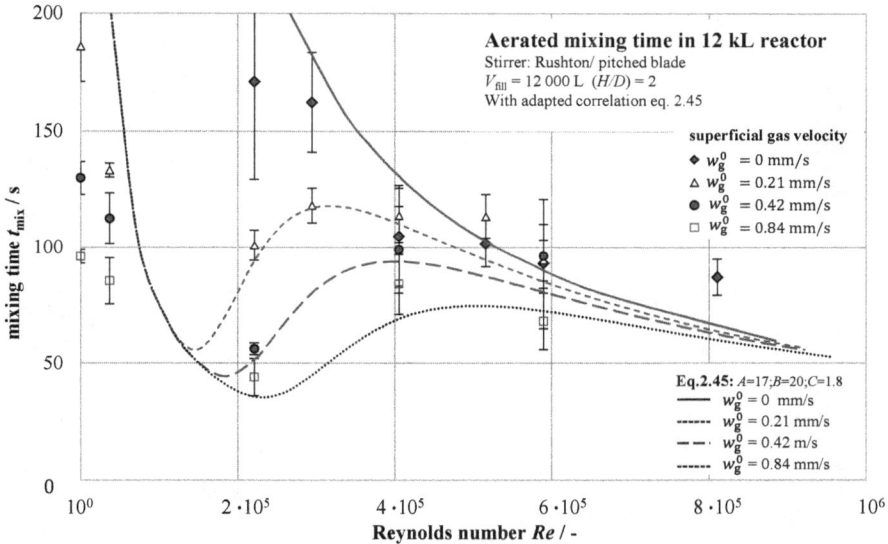

Figure 4-33: Mixing time as a function of the Reynolds number for different gassing rates [Ros18]

The behavior of the increase in mixing time with increasing power input can be explained as follows: with increasing stirrer frequency the momentum induced by the stirrer increases and leads to a better mixing within the reactor. The gassing rate also has a positive effect on the mixing, since the flow induced by buoyancy leads to large secondary flows and the break-up of stable liquid structures. Since the homogenisation leads to less large-scale vortices, the mixing performance decreases with increasing stirrer frequency in the range of lower Reynolds numbers. This operating behavior is of great importance for the pharmaceutical industry, as the volumetric power input for this application is limited. At higher Reynolds numbers the momentum induced by the Rushton turbine is dominant and the gas phase plays a minor role. Therefore, the mixing time under this condition is almost independent of the Reynolds number and the superficial gas velocity.

As can be seen in Figure 4-33, the correlation of Alves (eq.(2.44)) is in acceptable agreement with the experimental data. For this correlation, the critical Flow number was calculated using the equation by Mikulcova (eq. (2.20)) and the assumption that the influence of gassing on power input can be neglected ($P_g/P_0 = 1$). This has been previously confirmed by measurements of the power input. Nevertheless, it has to be mentioned that an adjustment of the parameters A, B and C of eq. (2.44) by experiments is necessary. The reliability of this correlation for different scales needs to be proved in more detail, especially when the parameters are kept constant.

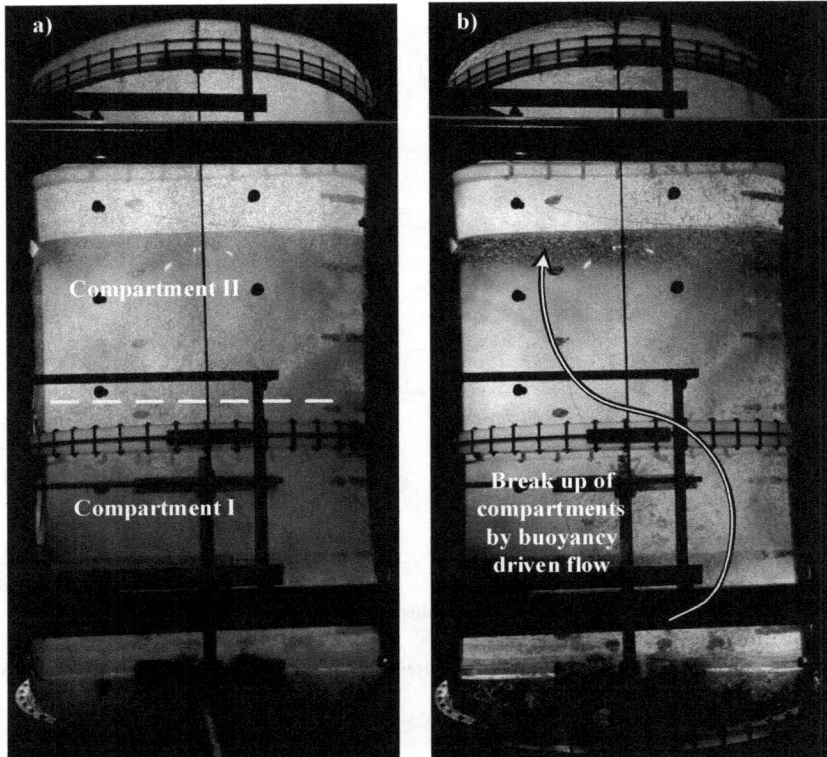

Figure 4-34: Comparison of the decolouration for two different operation conditions

a)
Operation condition:
$n = 80$ rpm
$w_G^0 = 0.4$ mm s^{-1}
$t = t_{mix} \cdot 0.5 = 50$ s
Fr $= 0.12/$ $Fl = 0.003$

b)
Operation condition:
$n = 30$ rpm
$w_G^0 = 0.4$ mm s^{-1}
$t = t_{mix} \cdot 0.5 = 26$ s
Fr $= 0.02/$ $Fl = 0.009$

The advantage of the total optical access is the possibility to visualize the mixing process in detail. In Figure 4-34 two decolouration processes are shown. Both pictures were taken after 50 % of the total mixing time. In Figure 4-34 a) it can be seen how the compartments are formed in the lower, well mixed part and the upper part of the reactor with a poor axial mixing. In this regime, the bubbles are rising homogeneously dispersed over the cross section of the reactor and are not able to break the barrier of the compartments, leading to long mixing times, especially in the upper part of the reactor. In Figure 4-34 b) the same superficial gas velocity but a decreased stirrer frequency had been adjusted. In this regime, the induced momentum by the impeller is insufficient to disperse and distribute the gaseous phase well over the cross section of the reactor, leading to a rise of large bubbles and thus to a strong buoyancy driven flow. This flow breaks up the barrier and leads to a better mass transfer to the upper region of the reactor which results in a shorter global mixing time. This effect can also be seen in the bubble size distribution. Figure 4-35 shows the heterogeneous regime with a very broad bubble size distribution and gas hold-up distribution whereas Figure 4-36 shows the same area at a higher Froude number and lower Flow number that causes a homogeneous flow regime with a narrow bubble size distribution as well as a homogeneous gas hold-up distribution over the reactor cross section.

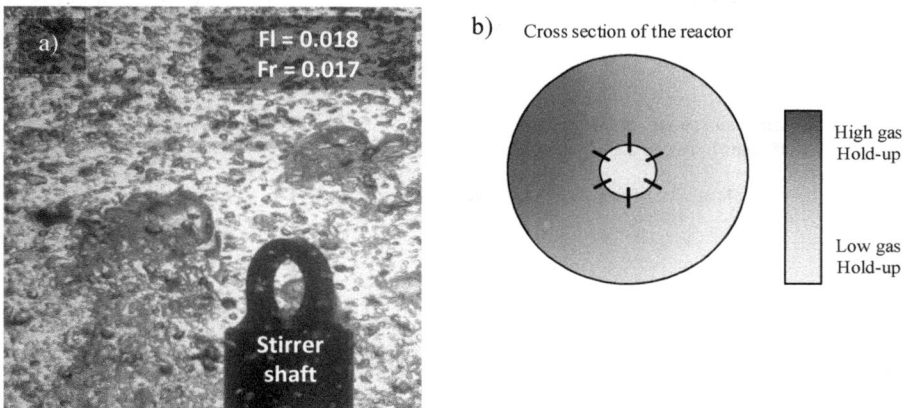

Figure 4-35:   Inhomogeneous bubbly flow – a) exemplary picture – b) inhomogeneous gas hold-up distribution

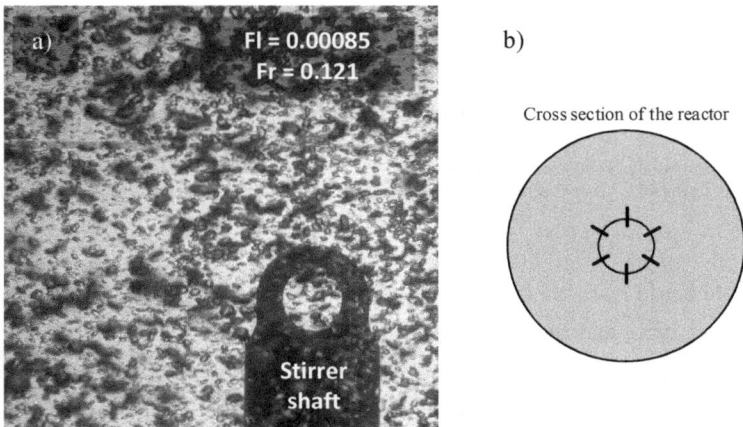

Figure 4-36:   Homogeneous bubbly flow – a) exemplary picture – b) homogeneous gas hold-up distribution

93

4.4   Mass transfer coefficient on lab scale and in industrial scale

The oxygen mass transfer is, beside of the mixing time, one of the most important parameter for aerobic processes and needs to be known for all used process scales.

In Figure 4-37, the volumetric mass transfer k_La coefficient is plotted as a function of the power input for two different superficial gas velocities (triangle and diamond) for the small scale system (red) and the large scale system (blue). The lines represent the correlation

$$k_La = C \cdot \left(\frac{P}{V}\right)^\alpha \cdot (w_G^0)^\beta \tag{4.1}.$$

For both systems the correlation could be fitted in order to allow a reliable prediction of the obtained results. The fitting parameters are listed in Table 4-1.

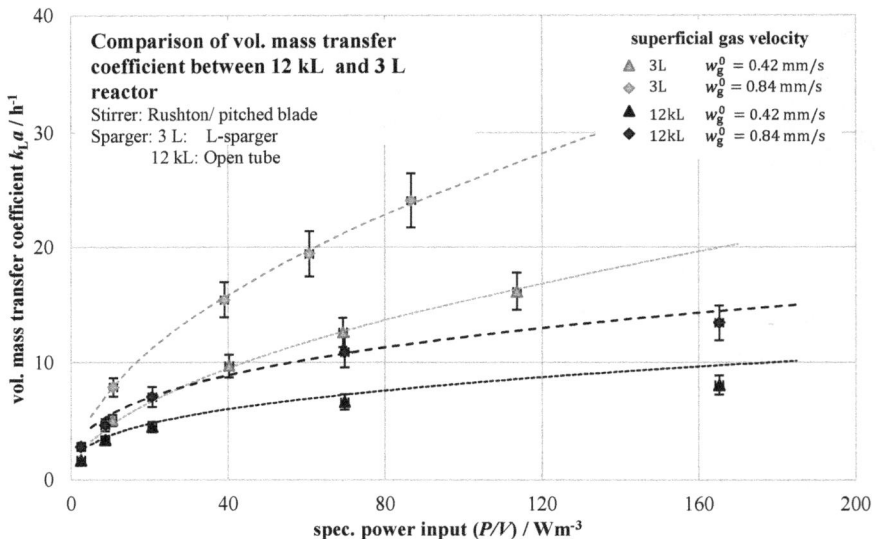

Figure 4-37:   Volumetric mass transfer coefficient as a function of the specific power input for two different superficial gas flow rates. Grey: values from small scale system. Black: values from large scale system

A comparison of the volumetric mass transfer coefficients in both systems generally shows that the values for the same power input and the same superficial gas velocities are of the same order of magnitude. However, it is noticeable that the power input in the small scale system has a higher influence on the mass transfer compared to the large scale system for the same superficial gas velocity. This trend can also be seen considering the fitting parameters α, β and C of the correlation. The values needed to be fitted separately for each scale but deliver a good determination coefficient R^2 and are listed in Table 4-1.

Table 4-1:   Fitting parameter for the correlation (4.1) for the small scale and the large scale system

	C	α	β	R^2
Small Scale	2.63	0.52	0.74	0.987
Large Scale	2.92	0.34	0.80	0.971

Comparison with literature shows that the values for small scale are similar to the ones shown in Table 2-8 in chapter 2.5.2. The values for α are given in literature for air/water systems in a range from 0.4 to 0.6. The values for β for these systems are in the range of 0.4 and 0.65. The found values for the small scale system are thus in good agreement to the proposed values.

The results in the large scale reactor on the other hand differ significantly from literature data that have been determined mainly in lab scale or pilot scale reactors. The influence of the power input with $\alpha = 0.34$ is considerably smaller, while the influence of the gas flow rate is stronger with $\beta = 0.80$.

The reason for the differences at the scales will be explained with further investigations by looking closer at the single variables k_L and a.

4.4.1   Mass transfer coefficient in small scale system

As described in chapter 2.5 the volumetric mass transfer coefficient is a combination of the mass transfer coefficient k_L and the specific surface area a. Therefore, the consideration of the combined factors in a detailed analysis of the mass transport has limitations. For instance, it cannot be distinguished whether the improved mass transport is caused by an increase in the mass transport coefficient or by a simple increase in the specific surface area. Thus, for a detailed analysis, the parameters need to be determined separately.

In the Figure 4-38 the determined mass transfer coefficient in the small scale system is plotted as a function of the power input for different superficial gas velocities. It is noticeable that the gas flow rate has almost no influence on the mass transfer coefficient. The differences that can be observed for higher gas flow rates are within the error bars and can thus be neglected. In comparison to the gas flow rate, the power input has a small influence on the mass transport coefficient. The mass transfer coefficient is increasing from $k_L = 0.88 \cdot 10^{-3}$ m/s at $(P/V) = 10$ W/m³ to $k_L = 1.55 \cdot 10^{-3}$ m/s at $(P/V) = 115$ W/m³, which is an increase of approximately 75 %. In the same range, however, the volumetric mass transfer coefficient k_La increased by more than 200 %. This leads to the assumption that the increase of the total volumetric mass transfer coefficient is mainly caused by a higher specific surface area.

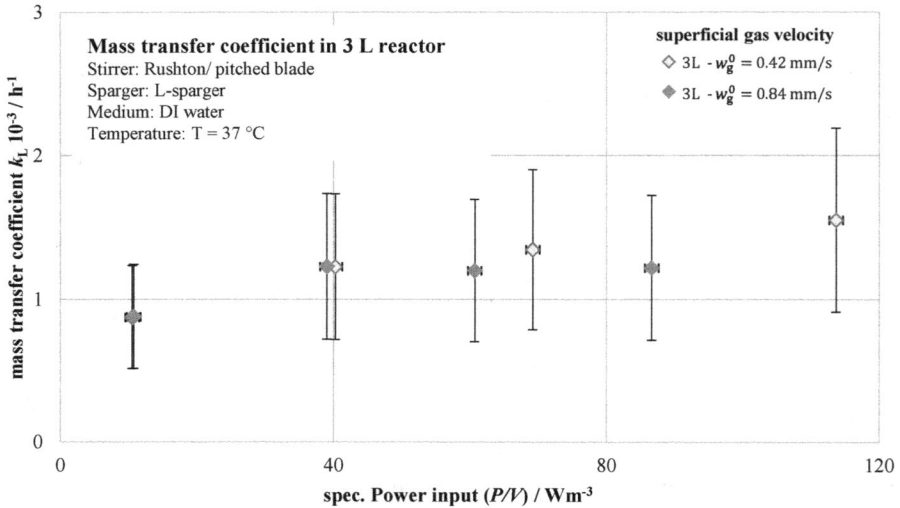

Figure 4-38: Mass transfer coefficient in 3 L system as a function of the specific power input for two different superficial gas velocities

In Figure 4-39, the experimental values for the mass transfer coefficient are compared with the "slip-velocity" model, which assumes that the renewal time is the time until the bubble has risen the length of the diameter, and with the "eddy" model, that assumes that the eddies present in the liquid leading to the surface renewal.

It can be seen that the correlation with the "eddy"- model shows a good agreement with the determined values. For the small superficial gas velocity, the calculated values are slightly below the determined values, but are clearly within the measurement uncertainties. Furthermore, they show the same trend with increasing specific power input. The values for the high superficial gas velocity are also in good agreement. However, the influence of the power input seems to be somewhat overestimated. Nevertheless, a similar trend can be observed. The "slip velocity"-model, assuming the bubble rise velocity as the main parameter, is underestimating the mass transfer coefficient in the whole investigation range. Furthermore, the influence of the power input cannot be predicted with this theory since the bubble sizes do not change in the same extend. Thus, the bubble rise velocity is not suitable for the calculation of the mass transfer. This comparison of the two theories shows that the mass transfer coefficient in the 3 L reactor depends mainly on the energy input. The bubble rise velocity only plays a minor role.

This conclusion gets confirmed for the small scale system when considering the Sherwood numbers. In Figure 4-40 the Sherwood numbers are compared with the correlation of Brauer and Mewes that has been developed for bubble columns with bubble sizes between $d_B = 3$ mm and $d_B = 10$ mm. In a bubble column it had been shown that the rise velocity is the main parameter for the mass transfer coefficient. The comparison clearly shows that the Sherwood number in the stirred tank reactor exceeds the values in the bubble column due to the influence of the stirred power input.

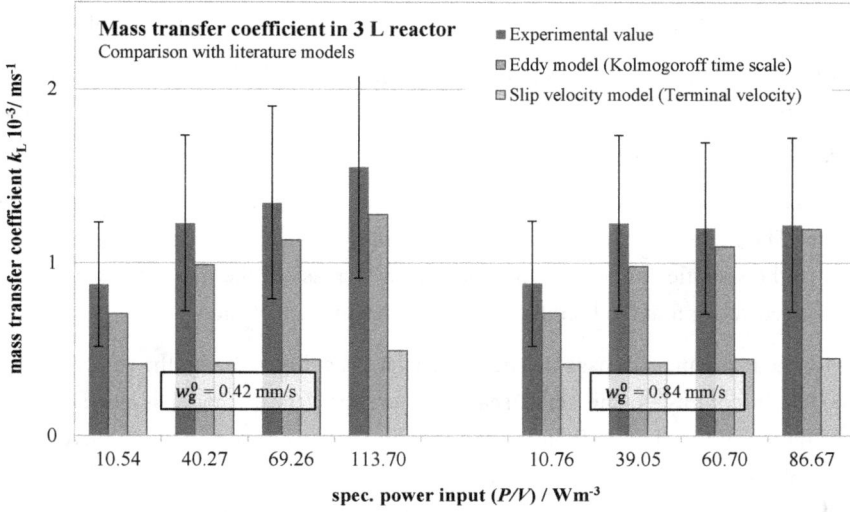

Figure 4-39: Comparison of the experimental mass transfer coefficient in 3 L reactor with correlations from literature

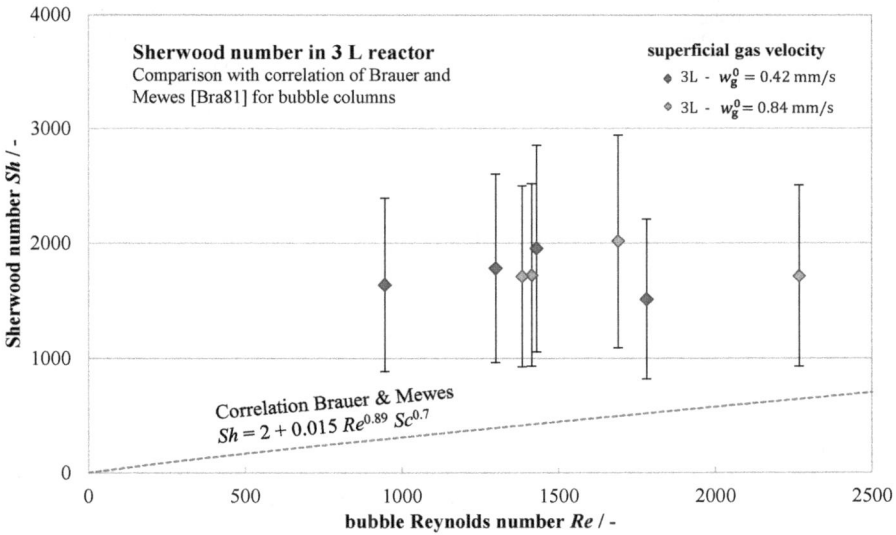

Figure 4-40: Comparison of the Sherwood number in the 3 L system and the bubble column correlation of Brauer and Mewes

4.4.2 Mass transfer coefficient in large scale system

The results for the calculated mass transfer coefficients in the 12 kL reactor are presented in Figure 4-41 as a funtion of the superficial gas velocity for two different specific power inputs.

In contrary to the small reactor, the mass transfer coefficients in the large scale reactor seem to be independent of both the gas flow rate and the power input. The change in total mass transfer is therefore only caused by the increase in surface area.

This leads to the conclusion that the bubbly flow in the large scale reactor behaves similar to a bubble column. The increased overall mass transfer with increasing power input is only caused by an expansion of the specific surface area and not by an increase of the mass transfer coefficient. A comparison with correlation of the literature comes to a similar conclusion.

In Figure 4-42, the experimental mass transfer coefficient is compared with the "eddy" and the "slip velocity" model. In contrast to the small scale reactor, the "eddy" model is not suitable for the estimation of the mass transfer coefficient. The influence of the power input is overestimated and does not fit to the experimental results. The "slip velocity" model, on the other hand, does only include the rise velocity and thus includes the power input only in terms of the bubble sizes. The error of the correlation is between 10 and 30 %. Taking into account the error bars, the results of the correlation lie within the deviations.

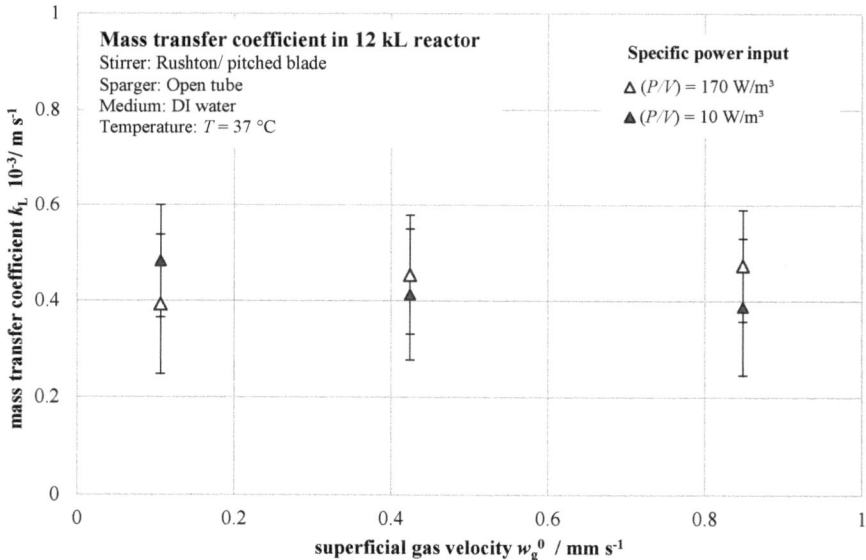

Figure 4-41: Mass transfer coefficient in 12 kL System as a function of the superficial gas velocities for two different specific power inputs

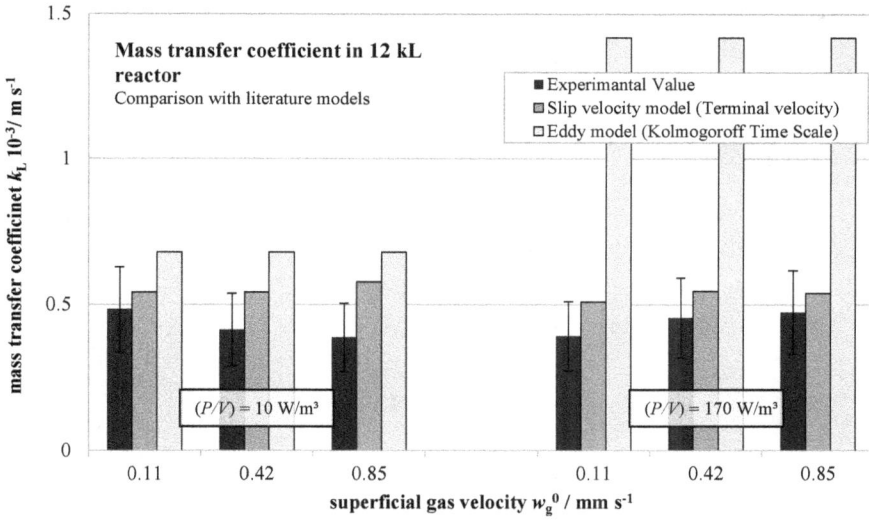

Figure 4-42: Comparison of the experimental mass transfer coefficient in 12 kL reactor and correlations from literature

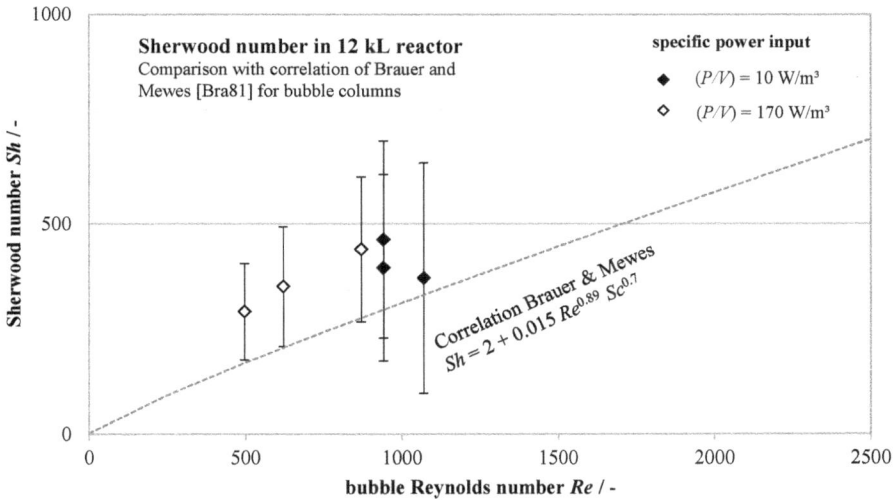

Figure 4-43: Comparison of the Sherwood number in the 12 kL system compared to the bubble column correlation of Brauer and Mewes

To confirm the theory, the calculated Sherwood numbers are compared with the correlation of Brauer and Mewes as well. The comparison can be seen in Figure 4-43. Contrary to the results from the 3 L reactor, the Sherwood numbers of the 12 kL reactor are in good agreement to the correlation.

This supports the thesis that in large reactors the influence of stirrer power on the mass transfer coefficient is reduced and that it is comparable to a bubble column.

4.4.3 Discussion

In the previous chapter, the mass transfer coefficients and their influencing parameters were investigated. It was found that, despite identical energy inputs and superficial gas velocities, the two scales differed greatly from each other. In order to determine the cause of these differences, the two scales are compared more precisely.

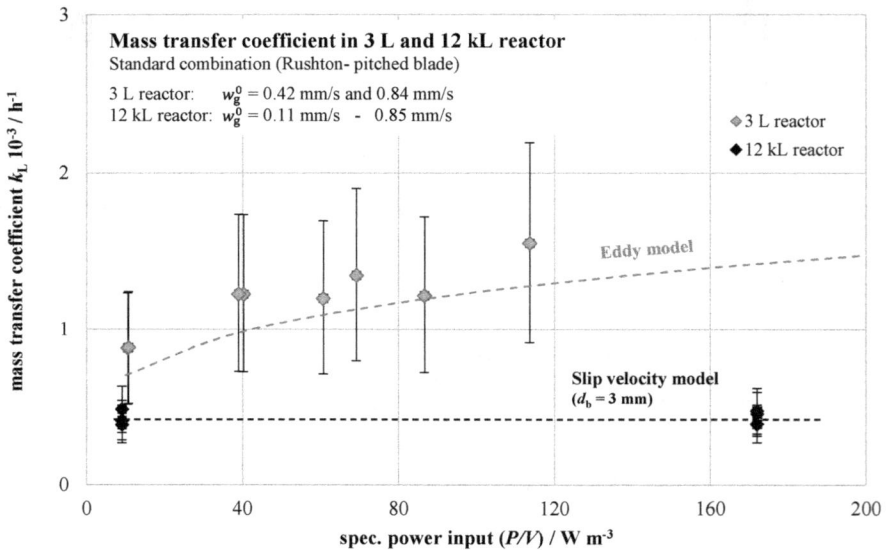

Mass transfer coefficient in 3 L and 12 kL reactor
Standard combination (Rushton- pitched blade)
3 L reactor: $w_g^0 = 0.42$ mm/s and 0.84 mm/s
12 kL reactor: $w_g^0 = 0.11$ mm/s - 0.85 mm/s

Eddy model

Slip velocity model
($d_b = 3$ mm)

◆ 3 L reactor
◆ 12 kL reactor

mass transfer coefficient k_L 10^{-3} / h^{-1}

spec. power input (P/V) / W m^{-3}

Figure 4-44: Mass transfer coefficient in small scale (red) and large scale (blue) system for different superficial gas velocities

In Figure 4-44, the mass transfer coefficient is plotted as a function of the power input for the small scale (red) and the large scale system (blue). It becomes clear that, on the one hand, the mass transfer coefficient in the small scale reactor is more than twice as large as that in the large scale reactor. On the other hand, the energy input in the 3 L reactor has a significant influence, while the mass transfer in the 12 000 L reactor remains almost constant. As already shown, the power input is not a sufficient parameter to describe the mass transfer coefficient in both scales.

There are different possible reasons for the deviations of the two scales.

1) False Measurement of the specific surface a:

In the large scale reactor, for the measurement of the bubble size distribution, only a small area was evaluated. Although this area has been carefully selected, errors in the determination of the specific surface area are not eliminated. Assuming that the evaluated mass transfer coefficient in the large scale reactor is too low, the specific surface area has to be smaller. This in turn means that the Sauter mean diameter has to be larger than determined. In order for the result to be in the same order of magnitude as in small scale, the Sauter mean diameter has to be twice as large. This, however, is very unlikely.

It is also possible that the specific surface area in the small scale reactor was determined too small, leading to an overestimated mass transfer coefficient. This may be due to the assumption of an elliptical bubble. Bubbles larger than $d_b = 5$ mm are having an irregularly form. If an ellipse is assumed, the determined surface is smaller than the actual one, which leads to a false measurement of the specific surface area. However, this assumption was made in both reactors and should therefore also lead to a wrong calculation in both reactors. It can therefore be assumed that there are actually differences in the mass transport mechanism.

2)   *Different influences of the small scale eddies at different scales*

The influence of the power input on the mass transport is based on the smallest eddies in the system. A higher energy input leads to smaller eddies, which increase the mass transport within the boundary layer. However, in order to affect the boundary layer, these eddies must be significantly smaller than the boundary layer. The size of the smallest eddies can be determined with $\eta_K = \left(\frac{\nu^3}{\varepsilon}\right)^{1/4}$. Since they depend only on the energy input, there is no explanation why they have a larger influence in the small scale system than in the large scale one. In addition, with the assumption of the local energy ε input to be 100 times higher in the impeller region compared to the average ε_T [Nie98], the smallest eddies are in the order of $\eta_K = 0.1$ mm at the highest power inputs. This is significantly larger than the boundary layer and it can be assumed that the power input cannot have a significant influence on the mass transfer coefficient k_L. Hence, another reason than eddies needs to be responsible for an increase of the mass transfer coefficient with increasing power input in the small system.

3)   *Different flow structures around the bubbles on different scales lead to different mass transfer coefficients*

   o   Pulsing of the bubbles, which cannot be detected with simple pictures

   o   Different shear at different scales lead to different flow behaviour and different interface detachment

When a bubble is exposed to shear rate due to velocity gradients, it will be deformed and the surface area will be enlarged (Figure 4-45). This additional surface area will be quickly enriched with oxygen. Especially when the bubble is deformed back into a sphere, the oxygen that was enriched in the extra-boundary layer is transported into the bulk phase (Figure 4-46). A high gradient flow thus leads to an increase in the overall mass transfer. On the other hand, a bubble that is approached with high velocity but is exposed to only a small gradient will follow the flow without much deformation. Thus, the mass transfer will not be increased significantly.

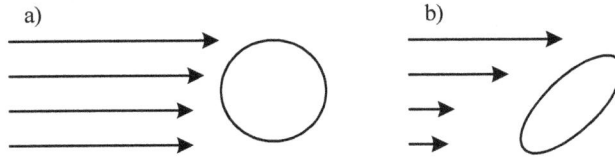

Figure 4-45: a) bubble exposed to high velocities but small velocity gradients b) bubble exposed to high velocity gradients

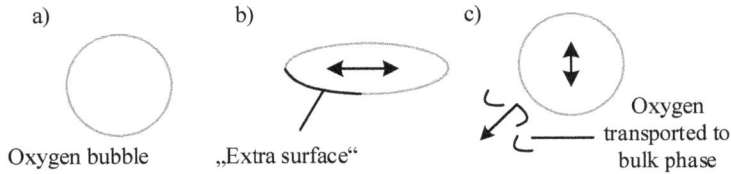

Figure 4-46: Mass transfer enhancement due to pulsing motion of an oxygen bubble

Since the power input and thus the flow pattern, but not the bubble sizes, are scaled, there is a significant difference in the shear of the bubbles in the different systems. The strongest shear rate in the bulk phase appears at the stirrer blade. As an example, the radial velocity can be considered as a function of the dimensionless axial position. Rousar [Rou94] found that the radial velocity u_r can be described with

$$\frac{u_r}{u_{tip}} = 0.78 \cdot \exp\left(-\frac{\left(2 * \frac{z}{h}\right)^2}{0.309}\right) \tag{4.2}$$

as a function of the axial position z and the stirrer height h.

With equation (4.2), the exact radial velocity can be determined for each process parameter with the tip speed u_{Tip}. In Table 4-2, both systems are compared with a low $((P/V)_{small} = 14$ W/m³ and $(P/V)_{large} = 10$ W/m³) and a high power input $((P/V) = 170$ W/m³).

In Figure 4-47, the calculated radial velocities are shown. The blue lines are the profiles in the large scale system; the red lines represent the velocities in the small scale system. The stirrer blades are also drawn in the correct size ratios. Due to the higher tip velocities in the large scale system, the maximum radial velocities are up to four times higher than in the small scale. However, it also becomes clear that the velocity gradients are significantly larger in the small system despite the smaller absolute velocity. For better comparison, a bubble with a diameter of $d_b = 5$ mm is also displayed in the correct size ratio. The velocity profile for the small scale system is additionally shown in the Figure 4-48, also with a bubble with $d_b = 5$ mm in the correct size.

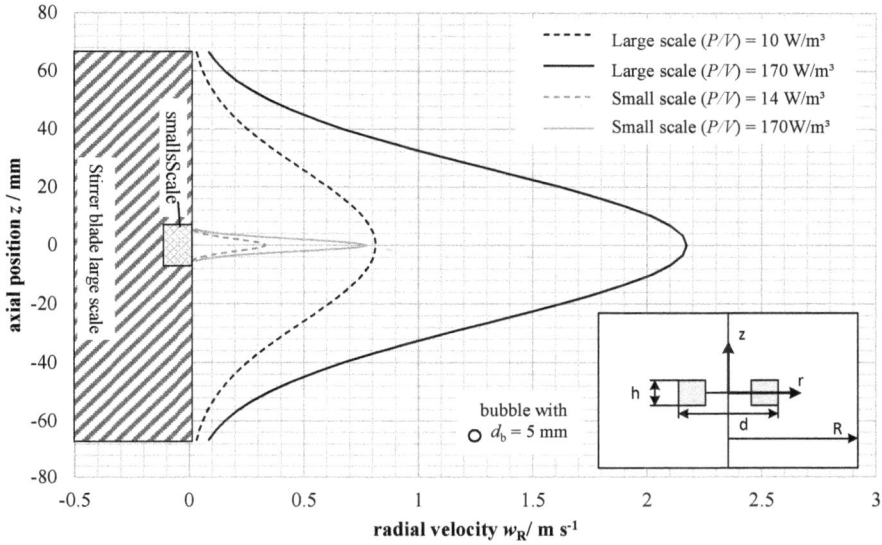

Figure 4-47: Radial velocity profile behind the stirrer blade at $r/R = 1/3$ for large scale (black) and small scale (grey)

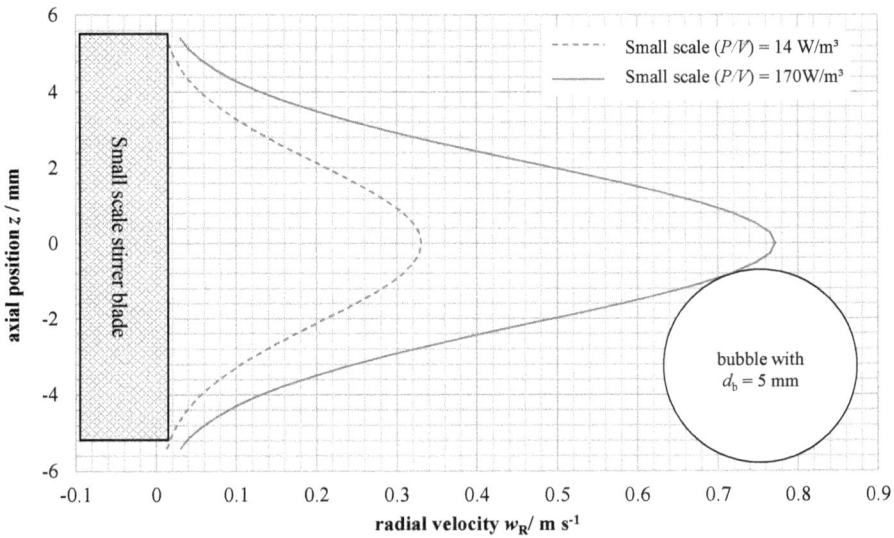

Figure 4-48: Radial velocity profile behind the stirrer blade at $r/R = 1/3$ in the small scale system

Table 4-2:   Dimensions and process parameters for small and large scale system

Small Scale Geometry	Process Parameter Small Power Input	High Power Input
$D = 0.15$ m	$(P/V) = 14$ W/m³	$(P/V) = 170$ W/m³
$d = 0.054$ m	$n = 150$ rpm	$n = 350$ rpm
$h = 0.011$ m	$u_{tip} = 0.424$ m/s	$u_{tip} = 0.990$ m/s
Large Scale Geometry		
$D = 2$ m	$(P/V) = 10$ W/m³	$(P/V) = 170$ W/m³
$d = 0.665$ m	$n = 30$ rpm	$n = 80$ rpm
$h = 0.133$ m	$u_{tip} = 0.424$ m/s	u_{tip} 2.786 m/s

An important measure is the shear stress $\tau = \eta \left(\frac{du}{dy}\right)$ that the bubbles and particles are exposed to. The calculated shear stress is presented in Figure 4-49. In this graph it becomes clear that the shear stress in the small scale system is up to five times larger than in the large system and thus leads to an increased shear and thus to larger interfacial surface areas.

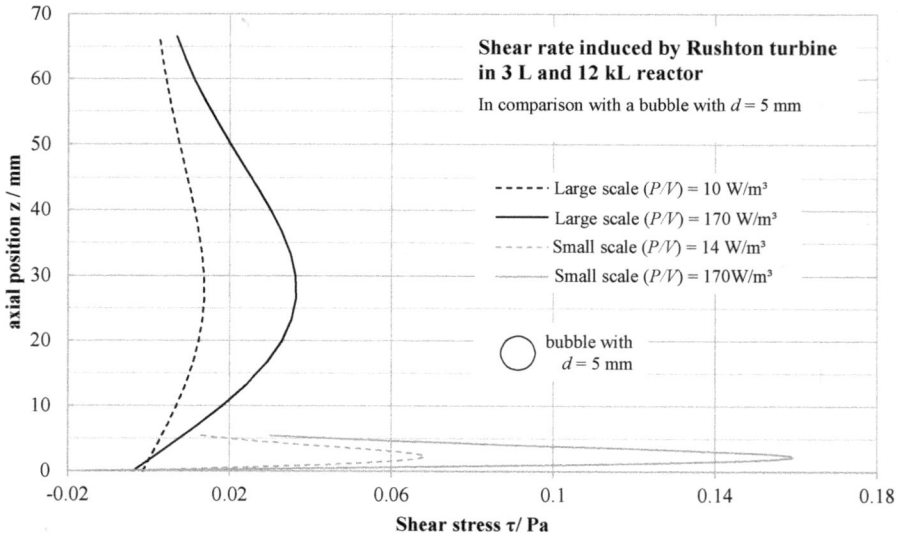

Figure 4-49:   Shear rate directely at the stirrer blade ($r/R = 1/3$)

This description can be applied to the entire reactor and it can be assumed that the shear induced by the velocity gradients tends to be larger in the small scale system. The different flows within the two scales can influence two important parameters for the mass transport coefficient. On the one hand, the mass transfer can be increased by a continuously changed surface of the bubbles, which is induced due to high shear rate. On the other hand, it has already been shown in many research projects on single bubbles that a changed flow can lead to a change of the mass transfer coefficient.

It has to be considered that the energy input in these investigations was relatively small and the resulting eddies remain rather large in relation to the boundary layer. In processes with significantly larger power inputs, the influencing parameters mentioned here can probably be negligible.

5. Modelling

With the investigations in the large scale reactor, the great influence of heterogeneity on the processes became clear. Both the mass transport and the mixing time were strongly affected by a change in the bubbly flow.

In the following chapter, the influence of heterogeneity on the two important process parameters will be presented and a modelling will be improved under consideration of the heterogeneity of the bubbly flow.

5.1   Mixing time

Precise knowledge of the mixing time is of great importance for a reliable process control. For single-phase systems, numerous correlations exist that predict the mixing time with sufficient accuracy. Usually, the dimensionless mixing time is used for these correlations. The influence of the gaseous phase is mostly described by the influence on the power input, which can lead to large errors, especially in the case of heterogeneous bubbly flows. While in single-phase systems the mixing time decreases with increasing stirrer frequency, in aerated processes a contrary behaviour can occur when the flow behaviour is changed from heterogeneous to homogeneous bubbly flow. This behaviour cannot be described by a simple consideration of the power input. Thus, the flow structure of the bubbly flow needs to be taken into account in the calculation of the aerated mixing time.

The investigation showed that over a wide range the gassing rate has a positive effect on the mixing time. This becomes also clear when for the same stirrer frequency the ratio of aerated to unaerated mixing time is plotted for each investigated point. The smaller the ratio, the more the mixing time is improved by the gassing. Mixing time ratios greater than one indicate that the mixing time has been increased. For ratios close to one, the aerated mixing time is in the same range as the unaerated and no significant influence of the gaseous phase can be assumed. In Figure 5-1, the mixing ratios for three aeration rates are plotted as a function of the Froude number. Due to high uncertainties of the measurement for the unaerated mixing time, the aerated mixing time is in some cases higher than the unaerated one. Nevertheless, it becomes clear that the aerated mixing time decreases with increasing gas flow rates. With increasing stirrer frequency, these differences become smaller until the ratio between the aerated mixing to unaerated mixing becomes one for all aeration rates. Furthermore, it can be seen that for higher aeration rates the stirrer frequency can be larger before the aerated/unaerated mixing time ratio becomes one. For instance at $Fr = 0.12$, the aerated/unaerated ratio is only smaller for the highest superficial gas velocity of $w_g^0 = 0.84$ mm/s. For $w_g^0 = 0.21$ mm/s and $w_g^0 = 0.42$ mm/s no differences between the aerated and the unaerated mixing time can be seen.

In chapter 4.3, it has been shown that the change of the flow regime has a strong influence on the mixing time. Chapter 4.2 displays that the change in the flow regime can be described with the critical Flow number Fl_c. A bubbly flow with a Flow number smaller than the critical Flow number can be assumed to be homogeneous. When the Flow number exceeds the critical Flow number the flow gets inhomogeneous with an increased influence of the gaseous phase on the mixing. Thus, the heterogeneity of a gaseous flow can be described with the ratio of the actual Flow number to the critical Flow number. A ratio below one describes a homogeneous flow regime. A ratio above one represents the heterogeneous flow regime.

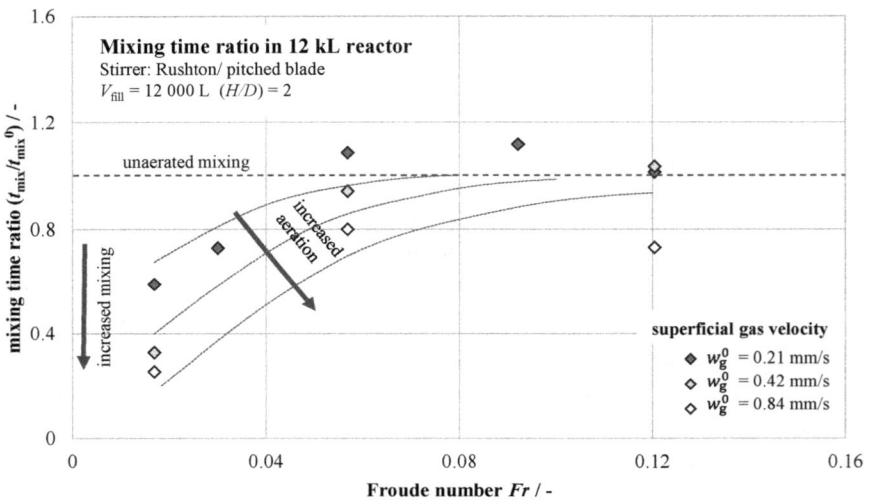

Figure 5-1: Mixing time ratio in dependency of the Froude number for three different superficial gas velocities

In Figure 5-2, the ratio of aerated to unaerated mixing time is plotted as a function of the heterogeneity of the gaseous flow (Fl_c/Fl). By plotting the mixing time ratio as a function of the heterogeneity, the differences between different gas flow rates disappear. The graph illustrates that the ratio between the aerated to the unaerated mixing time is increasing with increasing heterogeneity until the critical Flow number is reached. Reaching the critical flow number, i.e. with beginning of the heterogeneous flow regime, the ratio is not further decreased.

The trend can be estimated best with the correlation

$$\left(\frac{t_{mix}}{t_{mix}^0}\right) = \frac{1}{\left(a\left(\frac{Fl}{Fl_q}\right)+b\right)^2} + c \tag{5.1}$$

where a, b and c are fitting parameters.

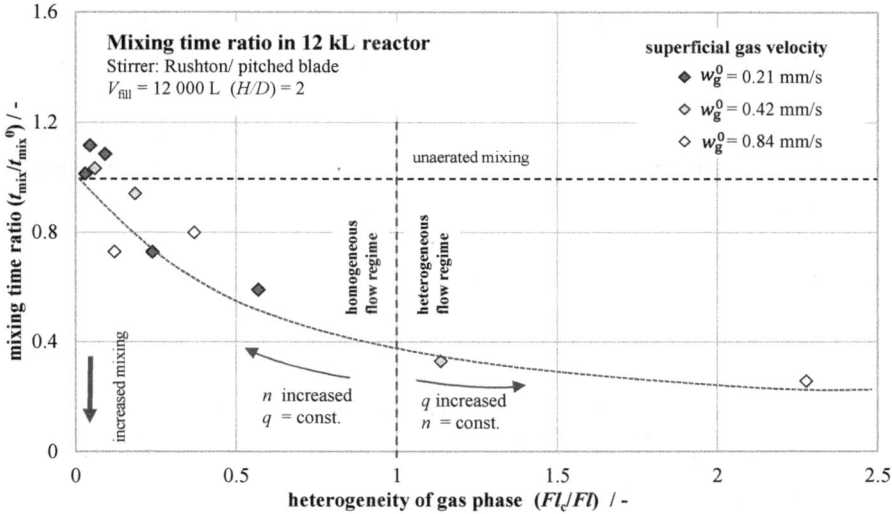

Figure 5-2: Mixing time ratio) in dependency of the heterogeneity for three different superficial gas velocities

The value a indicates how rapidly the maximum influence of the gassing is reached. A high value means that already small gassing rates are sufficient to have a large influence on the mixing time. The investigations at the acrylic glass reactor lead to a value $a = 0.8$.

The factor c gives the minimum aerated/unaerated factor and is dependent on the sparger. Spargers that provide good distribution of the gaseous phase even without the presence of an agitator, such as an aeration plate, lead to very homogeneous flows even at low stirrer speeds. Consequently, only a weak formation of a buoyancy driven flow takes place and thus, only a small influence of the gaseous phase on the mixing time can be assumed. For these agitators a high value of c can be expected. The investigated open tube sparger on the other hand, with a local and thus very heterogeneous aeration, leads to a strong influence on the mixing time and thus to a low value of c. The investigations at the 12 kL reactor with an open tube sparger lead to a minimum value of $c = 0.2$. This means that the aeration leads to a minimum decrease of $t_{mix} = 0.2 \cdot t_{mix}^0$.

The value b is a mathematical value which ensures that for a heterogeneity (Fl_c/Fl) approaching 0, the mixing ratio reaches $(t_{mix}/t_{mix}^0) = 1$ and is only linked to the value c with the formula

$$b = \left(\frac{1}{1-c}\right)^{0.5} \tag{5.2}.$$

The correlation with the fitted values for the acrylic glass reactor can be seen in Figure 5-3. For a better understanding of the fitting parameter and for a better comparability to literature values, additional experiments had been performed with other liquid heights and different stirrer combinations. The results for the level ratio of $(H/D) = 1$ with a Rushton turbine and a pitched blade (Rt-Pb), a level ratio of $(H/D) = 2.5$ with one Rushton turbine and two pitched blades (Rt-Pb-Pb) and the combination of two Ruston turbines (Rt-Rt) with $(H/D) = 2$ are included in Figure 5-3. It can be seen that correlation (5.1) is in good agreement to all three filling volumes with the combinations Rt-Pb and Rt-Pb-Pb. It has to be noted that the critical Flow number needed to be determined separately. They were found to be $Fl_c = 0.25 \cdot Fr$ for $(H/D) = 1$ and $Fl_c = 2 \cdot Fr$ for $(H/D) = 2.5$. The combination of two Rushton turbines, on the other hand, does not correspond well with the found correlation. The trends are similar, but in contrast to the combinations with one or two pitched blade turbines, the mixing time decreases at low Flow numbers. This means that the aerated to unaerated mixing time ratio is greater than one and does not correspond to the modelling. The increased mixing time with two Rushton turbines can be caused by a reduction of the power input during gassing. As this has not been investigated in this thesis, no conclusions can be made on this matter.

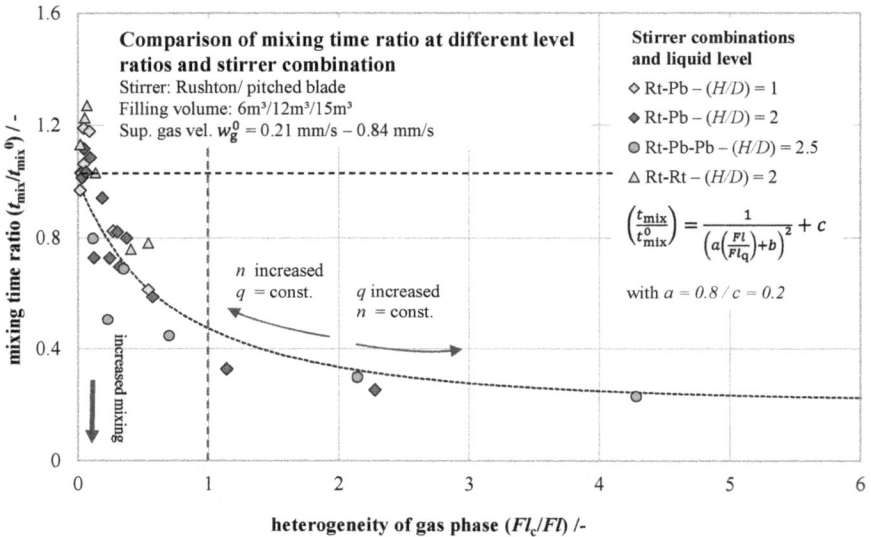

Figure 5-3: Correlation (5.1) in comparison with experimental results for four different stirrer combinations

With the new correlation the aerated mixing time for the standard combinations can be calculated within an error of ± 20 % (see Figure 5-4). For the same combination but with a level ratio of $(H/D) = 1$, the calculation is within an error of ± 30 % and for $(H/D) = 2.5$ within ± 36 %.

Adjusted values for the respective combination can be seen in Table 5-1. With the adjustment the error could be further reduced to 25 %. The adjustment of the values for a is reasonable, since the influence of the gassing rate does not have to be the same for different geometries or liquid heights. The table shows that the value a also increases with increasing filling volume. This means that for the same stirrer combination the influence of the Flow rate is increasing with increasing level ratio. This can be explained by the fact that the volumetric power input is not taken into account by the Flow number. For the same Flow number but double the filling volume the volumetric power input is only half. This means that a change of the Flow number has a larger influence on the axial flow and thus on the mixing time.

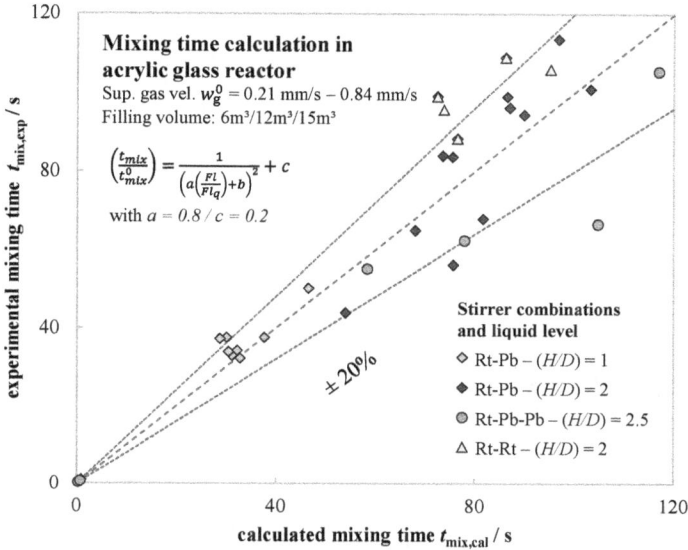

Figure 5-4: Comparison of correlated and measured mixing times for four different stirrer combinations

Table 5-1: Fitted parameter to the correlation (5.1) for the investigated stirrer combinations

Combination	Level ratio (H/D) /-	a	c	Max Error / %
Rt-Pb	1	0.5	0.2	25
	2	0.8	0.2	20
	2.5	1.5	0.2	25
Rt-Rt	2	0.3	0.2	30

For the Rushton-Rushton combination, the value a is the lowest of the investigated combinations. Due to the radial pumping of this turbine and the high power input, the gaseous phase does not have such a strong influence on the flow pattern compared to the axial pumping pitched blade turbine.

Comparison with literature

To be able to compare the correlation with literature data, information about the loading and flooding as well as about the mixing time must be available. Since investigations were usually carried out separately, only a limited amount of literature data is available. In Figure 5-5, data from Machon [Mac00] are presented. Machon investigated the mixing time in the transition regime of

111

loading and flooding. A reactor with a diameter $D = 0.29$ m with a Rushton turbine (RT), a down-pumping pitched blade turbine (PBTD), and an up-pumping pitched blade turbine (PBTU) had been used. Aeration was achieved by a ring sparger with six holes, located under the lower impeller.

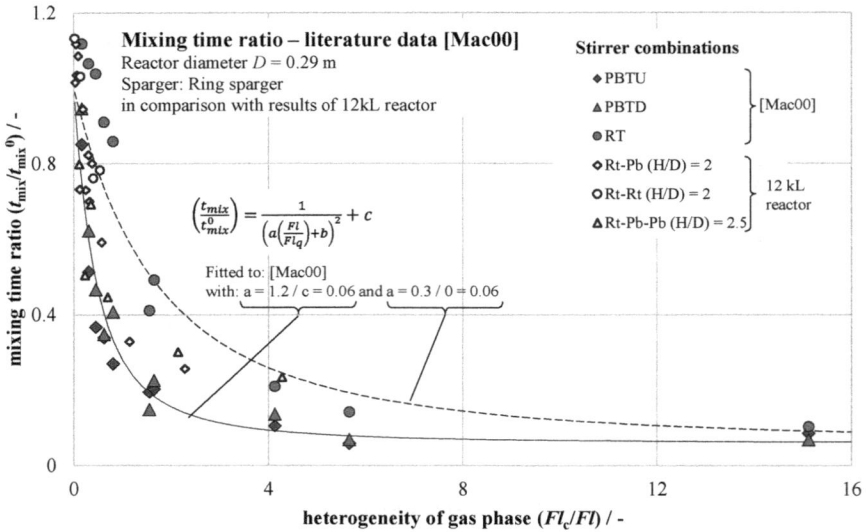

Figure 5-5: Correlation eq. (5.1) fitted to literature data [Mac00]

The plotting of the mixing time ratio against the heterogeneity shows that the same trend as in the acrylic glass reactor can be found in a significantly smaller reactor. It is also possible to apply the correlation with an error of $\pm$ 30 % for the pitched blade turbine and 50 % for the Rushton turbine combination. For the PBTU and the PBTD combinations, the same fitting parameters can be found. This means that similar flow structures occur in the reactor with the up-pumping and with the down-pumping pitched blade turbine. To fit the correlation to the results of the Rushton turbine combinations, a smaller value for a is found, because the mixing time ratio decreases much more slowly than for the pitched blade combinations. The value for c, however, is the same, because for the highest value of (Fl_c/Fl) the mixing time ratio reaches the minimum with $(t_{mix}/t_{mix}^0) = 0.06$. The found parameters are listed in Table 5-2.

In both reactors, the acrylic glass reactor and the reactor used by Machon [Mac00], the value of a for the Rushton turbine is noticeably lower than for the pitched blade combinations. Due to the good dispersion characteristics of the Rushton turbine, the gaseous phase has less influence on the flow structure and thus on the mixing time. Machon [Mac00] used a high reactor with a level ratio $(H/D) = 4$, thus the results are most likely to be compared with the results of $(H/D) = 2.5$ in the acrylic glass reactor. It can be seen that in the acrylic glass reactor, the aeration has a stronger influence ($a = 1.5$) compared to the smaller reactor ($a = 1.2$). This can be explained with the used open tube sparger, which provides a heterogeneous flow significantly earlier than a ring sparger.

Table 5-2: Fitted parameter to the correlation (5.1) for literature data [Mac00]

Combination	Level ratio (H/D) /-	a	c	Max Error / %
PBTU	4	1.2	0.06	30
PBTD	4	1.2	0.06	35
RT	4	0.3	0.06	50

The comparison with the literature data confirms the assumption that the value c depends on the type of sparger and reactor geometry and is a measure for the maximum reduction of the mixing time. For high gas flow rates the mixing time ratio is independent of the stirrer combination. The value a, on the other hand, depends on the stirrer geometry, the reactor geometry and the used sparger. A Rushton stirrer has a high power input and good dispersion characteristics. This leads to a slower reduction of the mixing time ratio compared to other stirrer types. Both, the results in the acrylic glass reactor and in the literature, show this behaviour. In both studies, the reduction of the mixing time is significantly slower when using a Rushton stirrer. Pitched blade stirrers, in contrast, have almost no dispersing effect and small gas volumes already lead to a considerable influence on the mixing time.

Furthermore, the type of sparger has a great influence on the mixing performance. An open tube was used in the acrylic gas reactor. This leads to a very strong buoyancy driven flow and thus to a strong axial mixing when the dispersion by the stirrer is not sufficient. With the same stirrer configuration, a ring sparger, which in itself ensures proper dispersion, will only have a significant influence on the mixing at higher Flow numbers.

With the help of the proposed correlation, the mixing time in the aerated condition can be determined while taking heterogeneity into account. Two fitting parameters and knowledge of the critical Flow number are necessary for the application. The critical Flow number can be determined by a simple measurement of the gas hold-up. For the determination of the values for a and c, further investigations in other systems are necessary. However, the two systems indicate that c depends solely on the geometry of the reactor and the aerator. The values for a are additionally dependent on the stirrer geometry. The values for Rushton turbines in the investigated systems with $a = 0.3$ are clearly below the values for pitched blade turbines. These were between $a = 0.5$ and $a = 1.5$, depending on the reactor height and the number of stirrers.

5.2   Mass transfer

The volumetric mass transport coefficient is one of the most important parameters for fermentation processes. Therefore, many studies on this topic have already been carried out and many correlations for its determination have been published. These correlations take into account geometric dimensions and the properties of the fluids. An overview was presented in chapter 2.5.2. However, the type of bubbly flow has not been considered yet.

In this study the volumetric mass transfer coefficient has been determined for the two different scales and the most common correlation has been fitted to the data (see chapter 4.4).

$$k_L a = C \cdot \left(\frac{P}{V}\right)^\alpha \cdot (w_g^0)^\beta \tag{5.3}$$

The results are listed again in the Table 5-3. It has been shown that the small scale results are in the same range as previously published fitting parameters. The results in large scale, in contrast, differ strongly from these values. For example, the influence of the power input on mass transfer is much smaller while the influence of the aeration is larger. Furthermore, the correlation could be fitted more suitably to the results in a small scale system than in the large reactor.

Table 5-3:   Fitting parameter of eq. (4.1) for the small and the large scale reactor

	C	α	β	R^2
Small Scale (3 L)	2.63	0.52	0.74	0.987
Large Scale (12 kL)	2.92	0.34	0.80	0.972

The largest difference between the two scales is the heterogeneous flow, which can only appear in large scale reactors. In the previous chapter it was shown that heterogeneity has a large influence on the mixing time. In order to determine the influence on the mass transfer coefficients, the deviation of the fitted correlation with the measured results will first be considered as a function of heterogeneity. The plot can be seen in Figure 5-6. It can be seen that the deviations for low heterogeneity are within a range of $\pm$ 15 %. For high heterogeneity the deviations are increasing to - 25 %. Thus, for high heterogeneities, the correlation generally underestimates the volumetric mass transfer coefficients, while it is correctly determined in the average for lower heterogeneities. Hence, it is reasonable to separate the heterogeneous and homogeneous flow results and to consider them separately.

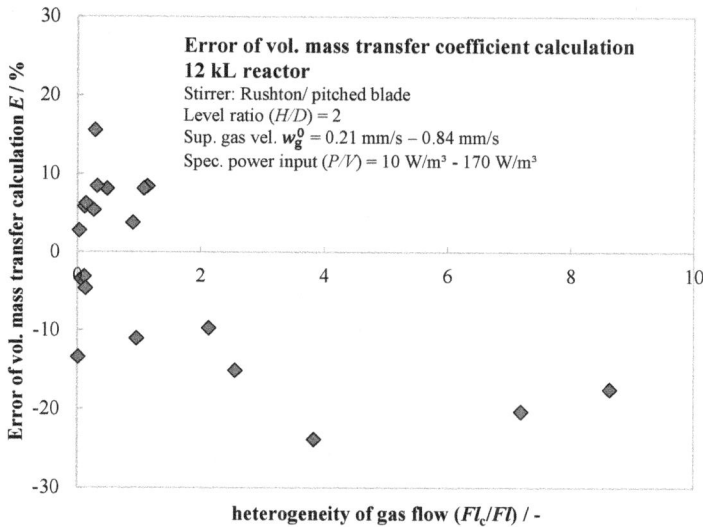

Figure 5-6: Error of calculation of volumetric mass transfer coefficient with eq. (5.3) in dependency of the heterogeneity

The fitting values for the separated flow regimes are listed in Table 5-4. First of all, it can be seen from the R^2 values that the results can be fitted significantly better when considered separately. When analysing the fitting parameters α and β, it becomes apparent how different the regimes are, also in terms of mass transport.

Table 5-4: Fitting parameter of eq. (4.1) for the small and the large scale reactor

	C	α	β	R^2
Small Scale	2.63	0.52	0.74	0.987
Large Scale - Total	2.92	0.34	0.80	0.972
Large Scale - homogeneous	4.87	0.24	0.83	0.991
Large Scale - heterogeneous	2.03	0.47	0.72	0.996

The fitted values confirm that the influences of the gassing rate and the energy input on the volumetric mass transfer coefficient within the two flow regimes differ significantly from each other. In the heterogeneous flow regime, the energy input has a much higher influence compared to the homogeneous flow regime, which is due to the fact that here the gas hold-up and thus the specific surface area is increased considerably more. The superficial gas velocity, on the other hand, has a stronger influence in the homogeneous flow regime, because an increase in superficial gas velocity in the heterogeneous flow regime leads to an even stronger inhomogeneity. As a result, large bubbles with small specific surface areas appear in the bubbly flow. The overall surface is enlarged, but with a much smaller factor than in the homogeneous flow regime.

In Figure 5-7, the two correlations for two superficial gas velocities are plotted as a function of the power input. Additionally, the experimental values are also added. It can be seen that in the homogeneous flow regime the power input only leads to a small change of the volumetric mass transfer coefficient. Only a variation in the superficial gas velocity leads to a significant change in mass transfer. If, however, the power input is reduced to such an extent that a heterogeneous flow is formed, a drastic drop in the mass transport capacity occurs.

The investigation showed that the volumetric mass transfer coefficient is strongly dependent on the flow regime. It is therefore recommendable to consider the regimes separately for the calculation. This increases the accuracy of the correlation from 25 % error for the total consideration to ± 10 % error for both the homogeneous and for the heterogeneous regime.

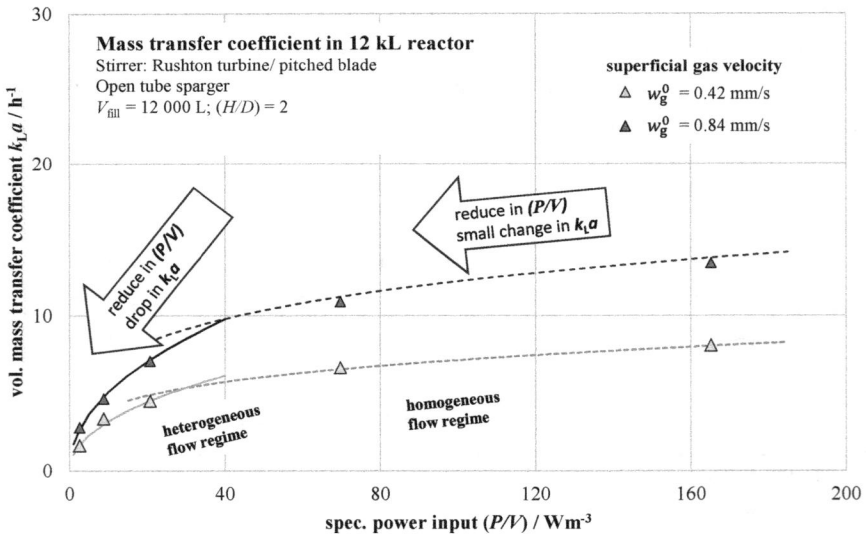

Figure 5-7: Correlation of the volumetric mass transfer coefficient in 12 kL reactor for two different superficial gas velocities for the heterogeneous (full lines) and the homogeneous flow regime (dotted lines)

6. Conclusion

At the new transparent acrylic glass reactor many new insights into the processes in large scale systems could be gained, which are not to be found in the small scale systems. Due to the small energy input in mammalian cell cultivation, the Kolmogoroff length scales are relatively large. This in turn means that even the small eddies are very large compared to the bubbles sizes, meaning that no direct influence is caused on the dispersion of the gaseous phase or the mass transport. As a consequence, the dispersion as well as the mass transport differed strongly from each other despite the same volumetric power input in the small scale and in the large scale system.

In the small scale system the dispersion mainly takes place at the sparger. Since the bubbles are large in relation to the vortices behind the blades, only limited further dispersion can be recognised in the impeller region. Instead, the gaseous phase accumulates behind the blades, forming large, gas-filled cavities. These lead to a significant reduction in the energy input and thus to a significant decrease in dispersion. On the large scale, on the other hand, the ratio of the vortices to the bubbles is much greater. This means that the gaseous phase is captured in these cavities and gets dispersed within them. Furthermore, since the bubbles are considerably smaller than the width of the stirrer blade, the formation of such gas filled cavities, as it can be seen in the small scale reactor, does not occur. Thus, hardly any difference in the power input can be seen in the large scale system under the investigated parameter.

Another important difference is the heterogeneity of the bubbly flow, which can be stronger in the large scale system compared to the small scale system. Due to the low power input, in contrast to microbial cultivation, the minimum speed may be exceeded at which sufficient dispersion occurs. Animal cell cultivation therefore often takes place in the transition area between homogeneous and heterogeneous flow. The flow has a particularly large influence on the mixing time. As a result, the correlations that exist for the aerated mixing time in the literature could not predict the trend sufficiently. Including the heterogeneity, which can be described by the critical flow number Fl_c, a new model could be developed in the context of this work, which reproduces the aerated mixing time satisfactorily, especially in the transition area. Also, literature data could be modelled with this correlation with good agreement. The heterogeneity also has a large influence on the mass transport, since especially the specific surface area is influenced. It has been shown that for the determination of the volumetric mass transfer coefficients, the two regimes have to be considered separately. The correlation could deliver better agreement with the experimental data.

The results of this work are a first step towards designing the laboratory reactors in such a way that they reproduce the processes in the production scale satisfactorily. Currently, both scales are operated with aerated stirred reactors with the same geometric ratio. However, it has been shown that many processes are fundamentally different due to the non-scalable bubble sizes. This refers to dispersion as well as to the heterogeneity of the entire bubble flow. In order to make the scales comparable, these aspects have to be considered more in the future.

Bibliography

[Aki73]   Akita, K.; Yoshida, F.: *Gas Holdup and Volumetric Mass Transfer Coefficient in Bubble Columns. Effects of Liquid Properties. In* Industrial & Engineering Chemistry Process Design and Development, 12; pp. 76–80, 1973.

[ALI81]   ALI, A. M. et al.: *Liquid Dispersion Mechanisms in Agitated Tanks. Part I. Pitched Blade Turbine. In* Chemical Engineering Communications, 10; pp. 205–213, 1981.

[Alv02]   Alves, S. S. et al.: *Bubble size in aerated stirred tanks. In* Chemical Engineering Journal, 89; pp. 109–117, 2002.

[Alv04]   Alves, S. S.; Maia, C. I.; Vasconcelos, J.: *Gas-liquid mass transfer coefficient in stirred tanks interpreted through bubble contamination kinetics. In* Chemical Engineering and Processing: Process Intensification, 43; pp. 823–830, 2004.

[Alv95]   Alves, S. S.; Vasconcelos, J. M.: *Mixing in gas-liquid contactors agitated by multiple turbines in the flooding regime. In* Chemical Engineering Science, 50; pp. 2355–2357, 1995.

[Arm99]   Armenante, P. M.; Mazzarotta, B.; Chang, G.-M.: *Power Consumption in Stirred Tanks Provided with Multiple Pitched-Blade Turbines. In* Industrial & Engineering Chemistry Research, 38; pp. 2809–2816, 1999.

[Asc15]   Ascanio, G.: *Mixing time in stirred vessels.* A review of experimental techniques. In Chinese Journal of Chemical Engineering, 23; pp. 1065–1076, 2015.

[Bae93]   Baehr, H.D and Stephan, K.: *Wärme- und Stofftransport.* Springer Berlin Heidelberg, 1993.

[Bat63]   Bates, R. L.; Fondy, P. L.; Corpstein, R. R.: *Examination of Some Geometric Parameters of Impeller Power. In* Industrial & Engineering Chemistry Process Design and Development, 2; pp. 310–314, 1963.

[BNC87] Bujalski, W. et al.: *The dependency on scale of power numbers of Rushton disc turbines. In* Chemical Engineering Science, 42; pp. 317–326, 1987.

[Bom06] Bombač, A.; Žun, I.: Individual impeller flooding in aerated vessel stirred by multiple-Rushton impellers. In Chemical Engineering Journal, 116; pp. 85–95, 2006.

[Bom97] Bombač, A. et al.: Gas-filled cavity structures and local void fraction distribution in aerated stirred vessel. In AIChE Journal, 43; pp. 2921–2931, 1997.

[Bot16] Bothe, M.: Experimental Analysis and Modeling of Industrial Two-Phase Flows in Bubble Column Reactors. Cuvillier Verlag, Hamburg, 2016.

[Bou01] Bouaifi, M. et al.: A comparative study of gas hold-up, bubble size, interfacial area and mass transfer coefficients in stirred gas–liquid reactors and bubble columns. In Chemical Engineering and Processing: Process Intensification, 40; pp. 97–111, 2001.

[Bou97] Bouaifi, M.; Roustan, M.; Djbbar, R.: *Hydrodynamics of Multi-Stage Agitated Gas-Liquid Reactors. In* Conference: 9th European Conference on Mixing, Paris, France, 1997.

[Bra71] Brauer, H.: *Grundlagen der Einphasen- und Mehrphasenströmungen.* Verlag Sauerländer, Aarau und Frankfurt am Mein, 1971.

[Bra81] Brauer, H.: *Particle/Fluid Transport Processes. In* Prog. Chem. eng, 19; pp. 61–99, 1981.

[Bus13] Busciglio, A. et al.: On the measurement of local gas hold-up, interfacial area and bubble size distribution in gas–liquid contactors via light sheet and image analysis. Imaging technique and experimental results. In Chemical Engineering Science, 102; pp. 551–566, 2013.

[Cal58] Calderbank, P.: Physical Rate Processes in Industrial Fermentation. Part 1. The Interfacial Area in Gas-liquid Contacting with Mechanical Agitation. In Trans. Inst. Chem. Eng., 36; pp. 443–463, 1958.

[Cam99] Camarasa, E. et al.: Influence of coalescence behaviour of the liquid and of gas sparging on hydrodynamics and bubble characteristics in a bubble column. In Chemical Engineering and Processing: Process Intensification, 38; pp. 329–344, 1999.

[CC88] Costes, J.; Couderc, J. P.: *Study by laser Doppler anemometry of the turbulent flow induced by a Rushton turbine in a stirred tank.* Influence of the size of the units—I. Mean flow and turbulence. In Chemical Engineering Science, 43; pp. 2751–2764, 1988.

[Cha16] Chara, Z. et al.: *Study of fluid flow in baffled vessels stirred by a Rushton standard impeller. In* Applied Mathematics and Computation, 272; pp. 614–628, 2016.

[CHA81] CHANG, T. et al.: Liquid Dispersion Mechanism in Agitated Tanks. Part II. Straight Blade and Disc Style Turbines. In Chemical Engineering Communications, 10; pp. 215–222, 1981.

[Che94] Chen, R. C.; Reese, J.; Fan, L.-S.: Flow structure in a three-dimensional bubble column and three-phase fluidized bed. In AIChE Journal, 40; pp. 1093–1104, 1994.

[Chr10] Christen, D. S.: *Praxiswissen der chemischen Verfahrenstechnik.* Handbuch für Chemiker und Verfahrensingenieure. Springer, Berlin, Heidelberg, 2010.

[Cli78] Clift, R.; Grace, J. R.; Weber, M. E.: *Bubbles, drops, and particles.* Dover Publ, Mineola, NY, 1987.

[Coo68] Cooper, R. G.; Wolf, D.: *Velocity profiles and pumping capacities for turbine type impellers. In* The Canadian Journal of Chemical Engineering, 46; pp. 94–100, 1968.

[Cut66] Cutter, L. A.: *Flow and turbulence in a stirred tank. In* AIChE Journal, 12; pp. 35–45, 1966.

[Dor13] Doran, P. M.: *Mixing.* In (Doran, P. M. Ed.): Bioprocess engineering principles. Academic Press, Waltham, MA; pp. 255–332, 2013.

[Fou44] Foust, H. C.; Mack, D. E.; Rushton, J. H.: *Gas-Liquid Contacting by Mixers. In* Industrial & Engineering Chemistry, 36; pp. 517–522, 1944.

[Fur12] Furukawa, H. et al.: *Correlation of Power Consumption for Several Kinds of Mixing Impellers. In* International Journal of Chemical Engineering, 2012; pp. 1–6, 2012.

[GAB08] Gill, N. K. et al.: Quantification of power consumption and oxygen transfer characteristics of a stirred miniature bioreactor for predictive fermentation scale-up. In Biotechnology and bioengineering, 100; pp. 1144–1155, 2008.

[Gab11] Gabelle, J.-C. et al.: *Effect of tank size on kLa and mixing time in aerated stirred reactors with non-newtonian fluids. In* The Canadian Journal of Chemical Engineering, 89; pp. 1139–1153, 2011.

[Gar04] Garcia-Ochoa, F.; Gomez, E.: Theoretical prediction of gas–liquid mass transfer coefficient, specific area and hold-up in sparged stirred tanks. In Chemical Engineering Science, 59; pp. 2489–2501, 2004.

[Gar09] Garcia-Ochoa, F.; Gomez, E.: Bioreactor scale-up and oxygen transfer rate in microbial processes: an overview. In Biotechnology advances, 27; pp. 153–176, 2009.

[Gra82]   Gray, D. J.; Treybal, R. E.; Barnett, S. M.: *Mixing of single and two phase systems.* Power consumption of impellers. In AIChE Journal, 28; pp. 195–199, 1982.

[Gre90]   Greaves, M.; Barigou, M.: *Estimation of Gas Hold - up and Impeller Power in a Stirred Vessel Reactor. In* "Fluid Mixing III", Inst. Chem. Engr., Int. Chem. Eng. Symp., 1990.

[Gui03]   Guillard, F.; Trägårdh, C.: *Mixing in industrial Rushton turbine-agitated reactors under aerated conditions. In* Chemical Engineering and Processing: Process Intensification, 42; pp. 373–386, 2003.

[Haß89]   Haß, V. C.; Nienow, A. W.: *Ein neuer, axial fördernder Rührer zum Dispergieren von Gas in Flüssigkeiten. In* Chemie Ingenieur Technik, 61; pp. 152–154, 1989.

[Hir01]   Hiraoka, S. et al.: *Power Consumption and Mixing Time in an Agitated Vessel with Double Impeller. In* Chemical Engineering Research and Design, 79; pp. 805–810, 2001.

[Hug67]   Hughmark, G. A.: *Holdup and Mass Transfer in Bubble Columns. In* Industrial & Engineering Chemistry Process Design and Development, 6; pp. 218–220, 1967.

[Hug80]   Hughmark, G. A.: *Power Requirements and Interfacial Area in Gas-Liquid Turbine Agitated Systems. In* Industrial & Engineering Chemistry Process Design and Development, 19; pp. 638–641, 1980.

[Inc02]   Incropera, F. P.; DeWitt, D. P.: *Fundamentals of heat and mass transfer.* J. Wiley, New York, 2002.

[Jav06]   Javed, K. H.; Mahmud, T.; Zhu, J. M.: *Numerical simulation of turbulent batch mixing in a vessel agitated by a Rushton turbine. In* Chemical Engineering and Processing: Process Intensification, 45; pp. 99–112, 2006.

[Jaw96]   Jaworski, Z.; Nienow, A. W.; Dyster, K. N.: An LDA study of the turbulent flow field in a baffled vessel agitated by an axial, down-pumping hydrofoil impeller. In The Canadian Journal of Chemical Engineering, 74; pp. 3–15, 1996.

[Jud76]   Judat, H.: Eignung schnellaufender Rührertypen zum Dispergieren von Gasen in niedrig-viskosen Flüssigkeiten. In Chemie Ingenieur Technik, 48; pp. 228–229, 1976.

[Kap06]   Kapic, A.; Heindel, T. J.: *Correlating Gas-Liquid Mass Transfer in a Stirred-Tank Reactor. In* Chemical Engineering Research and Design, 84; pp. 239–245, 2006.

[Kra12]   Kraume, M.: *Transportvorgänge in der Verfahrenstechnik*. Grundlagen und apparative Umsetzungen. Springer Berlin Heidelberg, Berlin, Heidelberg, 2012.

[Kri91]   Krishna, R.; Wilkinson, P. M.; van Dierendonck, L. L.: *A model for gas holdup in bubble columns incorporating the influence of gas density on flow regime transitions*. In Chemical Engineering Science, 46; pp. 2491–2496, 1991.

[KW91]   Kresta, S. M.; Wood, P. E.: *Prediction of the three-dimensional turbulent flow in stirred tanks*. In AIChE Journal, 37; pp. 448–460, 1991.

[KW93]   Kresta, S. M.; Wood, P. E.: *The mean flow field produced by a 45° pitched blade turbine*. Changes in the circulation pattern due to off bottom clearance. In The Canadian Journal of Chemical Engineering, 71; pp. 42–53, 1993.

[Lam70]   Lamont, J. C.; Scott, D. S.: An eddy cell model of mass transfer into the surface of a turbulent liquid. In AIChE Journal, 16; pp. 513–519, 1970.

[Lee14]   Lee, B. W.; Dudukovic, M. P.: *Determination of flow regime and gas holdup in gas–liquid stirred tanks*. In Chemical Engineering Science, 109; pp. 264–275, 2014.

[Lin04]   Linek, V. et al.: Gas–liquid mass transfer coefficient in stirred tanks interpreted through models of idealized eddy structure of turbulence in the bubble vicinity. In Chemical Engineering and Processing: Process Intensification, 43; pp. 1511–1517, 2004.

[Lin05a]   Linek, V.; Moucha, T.; Kordač, M.: *Mechanism of mass transfer from bubbles in dispersions*. In Chemical Engineering and Processing: Process Intensification, 44; pp. 353–361, 2005a.

[Lin05b]   Linek, V.; Kordač, M.; Moucha, T.: *Mechanism of mass transfer from bubbles in dispersions*. In Chemical Engineering and Processing: Process Intensification, 44; pp. 121–130, 2005b.

[Lu89]   Lu, W.-M.; Ju, S.-J.: Cavity configuration, flooding and pumping capacity of disc-type turbines in aerated stirred tanks. In Chemical Engineering Science, 44; pp. 333–342, 1989.

[Mac00]   Machon, V.; Jahoda, M.: *Liquid Homogenization in Aerated Multi-Impeller Stirred Vessel*. In Chemical Engineering & Technology, 23; pp. 869–876, 2000.

[Maj18]   Major-Godlewska, M.; Karcz, J.: Power consumption for an agitated vessel equipped with pitched blade turbine and short baffles. In Chemicke zvesti, 72; pp. 1081–1088, 2018.

123

[Mar08a] Martín, M.; Montes, F. J.; Galán, M. A.: *Bubbling process in stirred tank reactors II.* Agitator effect on the mass transfer rates. In Chemical Engineering Science, 63; pp. 3223–3234, 2008a.

[Mar08b] Martín, M.; Montes, F. J.; Galán, M. A.: *Bubbling process in stirred tank reactors I.* Agitator effect on bubble size, formation and rising. In Chemical Engineering Science, 63; pp. 3212–3222, 2008b.

[Mar09] Martín, M.; Montes, F. J.; Galán, M. A.: *Physical explanation of the empirical coefficients of gas–liquid mass transfer equations. In* Chemical Engineering Science, 64; pp. 410–425, 2009.

[McF96] McFarlane, C. M.; Nienow, A. W.: Studies of High Solidity Ratio Hydrofoil Impellers for Aerated Bioreactors. 3. Fluids of Enhanced Viscosity and Exhibiting Coalescence Repression. In Biotechnology Progress, 12; pp. 1–8, 1996.

[Mon08] Montante, G.; Horn, D.; Paglianti, A.: *Gas–liquid flow and bubble size distribution in stirred tanks. In* Chemical Engineering Science, 63; pp. 2107–2118, 2008.

[Mon15] Montante, G.; Paglianti, A.: *Gas hold-up distribution and mixing time in gas–liquid stirred tanks. In* Chemical Engineering Journal, 279; pp. 648–658, 2015.

[Mon99] Montes, F. J.; Galan, M. A.; Cerro, R. L.: *Mass transfer from oscillating bubbles in bioreactors. In* Chemical Engineering Science, 54; pp. 3127–3136, 1999.

[Mou03] Moucha, T.; Linek, V.; Prokopová, E.: *Gas hold-up, mixing time and gas–liquid volumetric mass transfer coefficient of various multiple-impeller configurations.* Rushton turbine, pitched blade and techmix impeller and their combinations. In Chemical Engineering Science, 58; pp. 1839–1846, 2003.

[Nie13] Nienow, A. W. et al.: The physical characterisation of a microscale parallel bioreactor platform with an industrial CHO cell line expressing an IgG4. In Biochemical Engineering Journal, 76; pp. 25–36, 2013.

[Nie71] Nienow, A. W.; Miles, D.: *Impeller Power Numbers in Closed Vessels. In* Industrial & Engineering Chemistry Process Design and Development, 10; pp. 41–43, 1971.

[Nie96] Nienow, A. W. et al.: Homogenisation and oxygen transfer rates in large agitated and sparged animal cell bioreactors: Some implications for growth and production. In Cytotechnology, 22; pp. 87–94, 1996.

[Nie97] Nienow, A. W.: On impeller circulation and mixing effectiveness in the turbulent flow regime. In Chemical Engineering Science, 52; pp. 2557–2565, 1997.

[Nie98]   Nienow, A. W.: *Hydrodynamics of Stirred Bioreactors. In* Applied Mechanics
           Reviews, 51; p. 3–3, 1998.

[OOK08]   Ochieng, A. et al.: *Mixing in a tank stirred by a Rushton turbine at a low
           clearance. In* Chemical Engineering and Processing: Process Intensification, 47;
           pp. 842–851, 2008.

[Pag02]   Paglianti, A.: Simple Model to Evaluate Loading/Flooding Transition in Aerated
           Vessels Stirred by Rushton Disc Turbines. In The Canadian Journal of Chemical
           Engineering, 80; pp. 1–5, 2002.

[Pau04]   Paul, E. L.; Atiemo-Obeng, V. A.; Kresta, S. M.: *Handbook of industrial mixing.*
           Science and practice. Wiley-Interscience, Hoboken N.J., 2004.

[Pee53]   Peebles, F. N.; Garber, H. J.: *Studies on the motion of gas bubble in liquids. In*
           Chem. Engng Prog. Symp. Ser, 49, 1953.

[Räb10]   Räbiger, N. et al.: *L4 Bubble and Drops in Technical Equipment:* VDI Heat
           Atlas. Springer Berlin Heidelberg, Berlin, Heidelberg; pp. 1239–1270, 2010.

[Reu70]   Reuß, M.: *Stoffübergang in Blasensäulen,* Berlin, 1970.

[Rew10]   Rewatkar, V. B.; Rao, K. R.; Joshi, J. B.: *Power Consumption in Mechanically
           Agitated Contactors using Pitched Blade Turbines Impellers. In* Chemical
           Engineering Communications, 88; pp. 69–90, 2010.

[Rew93]   Rewatkar, V. B. et al.: Gas hold-up behavior of mechanically agitated gas-liquid
           reactors using pitched blade downflow turbines. In The Canadian Journal of
           Chemical Engineering, 71; pp. 226–237, 1993.

[Ros18]   Rosseburg, A. et al.: Hydrodynamic inhomogeneities in large scale stirred tanks –
           Influence on mixing time. In Chemical Engineering Science, 188; pp. 208–220,
           2018.

[Rou94]   Rousar, I.; Van den akker, H. E. A.: LDA measurements of liquid velocities in
           sparged agitated tanks with single and multiple Rushton turbines. In In
           Proceedings of the 8th European Conference on Mixing, 1994.

[Rus68]   Rushton, J. H.; Bmbinet, J.-J.: *Holdup and flooding in air liquid mixing. In* The
           Canadian Journal of Chemical Engineering, 46; pp. 16–21, 1968.

[Sai92]   Saito, F. et al.: Power, gas dispersion and homogenisation characteristics of
           Scaba SRGT and rushton turbine impellers. In JOURNAL OF CHEMICAL
           ENGINEERING OF JAPAN, 25; pp. 281–287, 1992.

[SC95]   Stoots, C. M.; Calabrese, R. V.: *Mean velocity field relative to a Rushton turbine
           blade. In* AIChE Journal, 41; pp. 1–11, 1995.

[Sha82]   Shah, Y. T. et al.: *Design parameters estimations for bubble column reactors. In* AIChE Journal, 28; pp. 353–379, 1982.

[She06]   Shewale, S. D.; Pandit, A. B.: *Studies in multiple impeller agitated gas–liquid contactors. In* Chemical Engineering Science, 61; pp. 489–504, 2006.

[Sie16]   Sieblist, C.; Jenzsch, M.; Pohlscheidt, M.: Equipment characterization to mitigate risks during transfers of cell culture manufacturing processes. In Cytotechnology, 68; pp. 1381–1401, 2016.

[Smi04]   Smith, J. M.; Gao, Z.; Müller-Steinhagen, H.: *The effect of temperature on the void fraction in gas–liquid reactors. In* Experimental Thermal and Fluid Science, 28; pp. 473–478, 2004.

[Smi92]   Smith, J. M.: *Simple Performance Correlations for Agitated Vessels.* In (Moreau, R.; King, R. Eds.): Fluid Mechanics of Mixing. Modelling, Operations and Experimental Techniques. Springer Netherlands, Dordrecht; pp. 55–63, 1992.

[SVM77]   Smith, J.; Van't Riet, K.; Middelton, J.: *Scale - up of agitated gas - liquid reactors for mass transfer.* Proceedings of 2nd European Conference on Mixing, Cambridge, 1977.

[Tat80]   Tatterson, G. B.; Yuan, H.-H. S.; Brodkey, R. S.: *Stereoscopic visualization of the flows for pitched blade turbines. In* Chemical Engineering Science, 35; pp. 1369–1375, 1980.

[Tat91]   Tatterson, G. B.: *Fluid Mixing and Gas Dispersion in Agitated Tanks.* McGraw-Hill, New York, 1991.

[Van73]   Van't Riet, K.; Smith, J. M.: *The behaviour of gas—liquid mixtures near Rushton turbine blades. In* Chemical Engineering Science, 28; pp. 1031–1037, 1973.

[Van79]   Van't Riet, K.: *Review of Measuring Methods and Results in Nonviscous Gas-Liquid Mass Transfer in Stirred Vessels. In* Industrial & Engineering Chemistry Process Design and Development, 18; pp. 357–364, 1979.

[Vas95]   Vasconcelos, J. M.; Alves, S. S.; Barata, J. M.: *Mixing in gas-liquid contactors agitated by multiple turbines. In* Chemical Engineering Science, 50; pp. 2343–2354, 1995.

[VBF00]   Vilaça, P. R. et al.: Determination of power consumption and volumetric oxygen transfer coefficient in bioreactors. In Bioprocess Engineering, 22; pp. 261–265, 2000.

[Vrá00]   Vrábel, P. et al.: *Mixing in large-scale vessels stirred with multiple radial or radial and axial up-pumping impellers.* Modelling and measurements. In Chemical Engineering Science, 55; pp. 5881–5896, 2000.

[Vrá99]   Vrábel, P. et al.: *Compartment Model Approach.* In Chemical Engineering Research and Design, 77; pp. 291–302, 1999.

[VS75]   Van't Riet, K.; Smith, J. M.: *The trailing vortex system produced by Rushton turbine agitators.* In Chemical Engineering Science, 30; pp. 1093–1105, 1975.

[Wan13]   Wang, D.; Fan, L.-S.: *Particle characterization and behavior relevant to fluidized bed combustion and gasification systems.* In (Scala, F. Ed.): Fluidized-bed technologies for near-zero emission combustion and gasification. Woodhead publ, Oxford; pp. 42–76, op. 2013.

[War85]   Warmoeskerken, M.; Smith, J. M.: *Flooding of disc turbines in gas-liquid dispersions.* A new description of the phenomenon. In Chemical Engineering Science, 40; pp. 2063–2071, 1985.

[Wie83]   Wiedmann, J. A.: Zum überflutungsverhalten zwei- und dreiphasig betriebener Rührreaktoren. In Chemie Ingenieur Technik, 55; pp. 689–700, 1983.

[Wu89]   Wu, H.; Patterson, G. K.: *Laser-Doppler measurements of turbulent-flow parameters in a stirred mixer.* In Chemical Engineering Science, 44; pp. 2207–2221, 1989.

[Yaw02]   Yawalkar, A. A.; Pangarkar, V. G.; Beenackers, A. A. C. M.: *Gas hold-up in stirred tank reactors.* In The Canadian Journal of Chemical Engineering, 80; pp. 158–166, 2002.

[Yos63]   Yoshida, F.; Miura, Y.: *Gas Absorption in Agitated Gas-Liquid Contactors.* In Industrial & Engineering Chemistry Process Design and Development, 2; pp. 263–268, 1963.

[Zah96]   Zahradník, J.; Fialová, M.: The effect of bubbling regime on gas and liquid phase mixing in bubble column reactors. In Chemical Engineering Science, 51; pp. 2491–2500, 1996.

[ZBW01]   Zhu, Y.; Bandopadhayay, P. C.; Wu, J. I.: *Measurement of Gas-Liquid Mass Transfer in an Agitated Vessel. A Comparison between Different Impellers.* In Journal of Chemical Engeneering of Japan, 34; pp. 579–584, 2001.

[Zlo03]   Zlokarnik, M.: *Stirring.* Wiley - VCH Verlag GmbH & Co. KGaA, 2003.

[Zlo67a] Zlokarnik, M.; Judat, H.: *Rohr- und Scheibenrührer - zwei leistungsfähige Rührer zur Flüssigkeitsbegasung. In* Chemie Ingenieur Technik, 39; pp. 1163–1168, 1967a.

[Zlo67b] Zlokarnik, M.: *Eignung von Rührern zum Homogenisieren von Flüssigkeitsgemischen. In* Chemie Ingenieur Technik, 39; pp. 539–548, 1967b.

[Zlo73] Zlokarnik, M.: *Rührleistung in begasten Flüssigkeiten. In* Chemie Ingenieur Technik, 45; pp. 689–692, 1973.

Supervised Thesis

"Experimentelle Analyse des Stoffdurchganges zweiphasiger Systeme in begasten Rührkesselreaktoren" *J. Fitschen, Juli 2017*

Entwicklung und Implementierung einer Messmethode zur Bestimmung des Stofftransportkoeffizienten unter stationären Bedingungen" *S. Nichtern, September 2017*

„Gegenüberstellung unterschiedlicher Methoden zur Ermittlung der spezifischen Phasengrenzfläche im Labor- und Industriemaßstab", *V. Berg, September 2016*

„Bestimmung und Vergleich der Blasengrößenverteilung in begasten Rührreaktoren im Labor- und Industriemaßstab unter Variation wichtiger Prozessparameter", *D. Sanders, September 2016*

„Untersuchung des Einflusses von blasenzerteilenden Gittern sowie unregelmäßigem Rühren auf den Stofftransport in begasten Rührreaktoren", *T. Kuczynski, June 2016*

„Optimierung und Anwendung der Auswertemethodik zur Bestimmung des volumetrischen Stoffübergangskoeffizienten in begasten Rührreaktionen", *B. Niclas, June 2016*

„Untersuchungen zu Phasengrenzfläche und Stofftransport für verschiedene Konfigurationen eines Single-Use Bioreaktors", *M. Maly, November 2015*

„Untersuchung des Einflusses von unterschiedlichen Rührwerken und Begasungsraten auf die Mischzeit eines 2 L Rührreaktors", *K. Sadat, October 2015*

„Xanthan Gum aqueous solution: Rheological properties and influence of non-ionic surfactant addition", *A. Miravalles, April 2014*

„Rheologische Charakterisierung von Fermentationsbrühen zur Entwicklung von Modellfluiden", *C. Kloock, December 2013*

„Bestimmung der Blasengrößenverteilungin Modellmedien von Fermentationsbrühen", *L. Parakenings, Oktober 2013*

Lebenslauf

Name:	Rosseburg
Vorname:	Annika
Staatsangehörigkeit:	deutsch
Geburtsdatum:	17.08.1986
Geburtsort:	Mannheim, Deutschland

08.1993 - 07.1997	Grundschule in Calberlah
08.1997 - 07.1999	Orientierungsstufe Isenbüttel
08.1999 - 07.2006	Otto-Hahn Gymnasium, Gifhorn
10.2006 - 04.2013	Studium Verfahrenstechnik an der Technischen Universität Hamburg (TUHH) Abschluss: Diplom
06.2013 - 12.2018	Wissenschaftliche Mitarbeiterin an der Technischen Universität Hamburg, Institut für Mehrphasenströmungen
01.2019 - heute	Anfertigung der Dissertation an der Technischen Universität Hamburg, Institut für Mehrphasenströmungen

List of Publications

Journal articles

Rosseburg, A.; Fitschen, J.; Wutz, J.; Wucherpfennig, T.; Schlüter, M.: Hydrodynamic inhomogeneities in large scale stirred tanks – Influence on mixing time,Chemical Engineering Science, 2018, 188, 208-220, DOI: 10.1016/j.ces.2018.05.008

Oral presentation

Rosseburg, A., Fitschen, J., Wutz, J., Wucherpfennig, T., Schlüter, M., "Heterogeneities in Large Scale Stirred Tank Reactors - new insights into a black box", Köthener Rührer-Kolloquium, Köthen 2018, Germany

Rosseburg, A., Fitschen, J., Wutz, J., Wucherpfennig, T., Schlüter, M., „Influence of local flow structure on the mixing behaviour within an aerated 12 000L stirred tank reactor", Jahrestreffen der ProcessNet Fachgruppe Mischvorgänge, Munich 2018, Germany

Rosseburg,A.; Schäfer, J.-E.; Wucherpfennig,T.; Berger,M.; Schlüter, M. "Investigation of local mixing and bubble size distribution in a stirred bubble column reactor", 9th International Conference on Multiphase Flows, Firenze 2016, Italy

Poster presentation

Rosseburg, A., Fitschen, J., Wutz, J., Wucherpfennig, T., Schlüter, M „Influence of local flow structure on the mixing behaviour within an aerated 12 000L stirred tank reactor", Himmelfahrtstagung, Magdeburg 2018, Germany